Mathematics:
Its Historical Aspects,
Wonders and Beyond

Problem Solving in Mathematics and Beyond

Print ISSN: 2591-7234
Online ISSN: 2591-7242

Series Editor: Dr. Alfred S. Posamentier
Distinguished Lecturer
New York City College of Technology - City University of New York

There are countless applications that would be considered problem solving in mathematics and beyond. One could even argue that most of mathematics in one way or another involves solving problems. However, this series is intended to be of interest to the general audience with the sole purpose of demonstrating the power and beauty of mathematics through clever problem-solving experiences.

Each of the books will be aimed at the general audience, which implies that the writing level will be such that it will not engulfed in technical language — rather the language will be simple everyday language so that the focus can remain on the content and not be distracted by unnecessarily sophiscated language. Again, the primary purpose of this series is to approach the topic of mathematics problem-solving in a most appealing and attractive way in order to win more of the general public to appreciate his most important subject rather than to fear it. At the same time we expect that professionals in the scientific community will also find these books attractive, as they will provide many entertaining surprises for the unsuspecting reader.

Published

For the complete list of volumes in this series, please visit www.worldscientific.com/series/psmb

Problem Solving in
Mathematics and Beyond

Volume 27

Mathematics:
Its Historical Aspects, Wonders and Beyond

Alfred S. Posamentier
The City University of New York, USA

Arthur D. Kramer
The City University of New York, USA

 World Scientific

NEW JERSEY · LONDON · SINGAPORE · BEIJING · SHANGHAI · HONG KONG · TAIPEI · CHENNAI · TOKYO

Published by

World Scientific Publishing Co. Pte. Ltd.

5 Toh Tuck Link, Singapore 596224

USA office: 27 Warren Street, Suite 401-402, Hackensack, NJ 07601

UK office: 57 Shelton Street, Covent Garden, London WC2H 9HE

Library of Congress Cataloging-in-Publication Data

Names: Posamentier, Alfred S., author. | Kramer, Arthur D., 1940–   author.

Title: Mathematics : its historical aspects, wonders and beyond /
   Alfred S. Posamentier, The City University of New York, USA, and
   Arthur D. Kramer, The City University of New York, USA.

Description: New Jersey : World Scientific, [2022] |
   Series: Problem solving in mathematics and beyond, 2591-7234 ; 27 | Includes index.

Identifiers: LCCN 2022019349 | ISBN 9789811248573 (hardcover) |
   ISBN 9789811249334 (paperback) | ISBN 9789811248580 (ebook) |
   ISBN 9789811248597 (ebook other)

Subjects: LCSH: Mathematics--History

Classification: LCC QA21 .P67 2022 | DDC 510--dc23/eng/20220505

LC record available at https://lccn.loc.gov/2022019349

British Library Cataloguing-in-Publication Data

A catalogue record for this book is available from the British Library.

For any available supplementary material, please visit
https://www.worldscientific.com/worldscibooks/10.1142/12594#t=suppl

Desk Editors: Vishnu Mohan/Rok Ting Tan/Qifan Tan

Typeset by Stallion Press
Email: enquiries@stallionpress.com

Printed in Singapore

About the Authors

Alfred S. Posamentier is currently Distinguished Lecturer at New York City College of Technology of the City University of New York. Previously, he was the Executive Director for Internationalization and Sponsored Programs at Long Island University, New York. This was preceded by a five-year period where, he was Dean of the School of Education and tenured Professor of Mathematics Education at Mercy College, New York. He is now also Professor Emeritus of Mathematics Education at The City College of the City University of New York, and former Dean of the School of Education, where he was tenured for 40 years. He is the author and co-author of more than 60 mathematics books for teachers, secondary and elementary school students, and the general readership. Dr. Posamentier is also a frequent commentator in newspapers and journals on topics relating to education. After completing his B.A. degree in mathematics at Hunter College of the City University of New York in 1964, he took a position as a teacher of mathematics at Theodore Roosevelt High School (Bronx, New York), where he focused his attention on improving the students' problem-solving skills and at the same time enriching their instruction far beyond what the traditional textbooks offered. During his six-year tenure there, he also developed the school's first mathematics teams (both at the junior and senior level). He is still involved in working with mathematics teachers and supervisors, nationally and internationally, to help them maximize their effectiveness. During this time, he earned an M.A. degree at the City College of the City University of New York in 1966.

Immediately upon joining the faculty of the City College in 1970, he began to develop in-service courses for secondary school mathematics teachers, including such special areas as recreational mathematics and problem-solving in mathematics. As Dean of the City College School of Education for 10 years, his scope of interest in educational issues covered the full gamut educational issues. During his tenure as dean he took the School from the bottom of the New York State rankings to the top with a perfect NCATE accreditation assessment in 2009. He achieved the same success in 2014 at Mercy College, which received both NCATE and CAEP accreditation during his leadership as dean of the School of Education.

In 1973, Dr. Posamentier received his Ph.D. from Fordham University (New York) in mathematics education and has since extended his reputation in mathematics education to Europe. He has been visiting professor at several European universities in Austria, England, Germany, Czech Republic, and Poland, while at the University of Vienna he was Fulbright Professor (1990). In 1989 he was awarded an Honorary Fellow at the South Bank University (London, England). In recognition of his outstanding teaching, the City College Alumni Association named him Educator of the Year in 1994, and in 2009. New York City had the day, May 1, 1994, named in his honor by the President of the New York City Council. In 1994, he was also awarded the Grand Medal of Honor from the Republic of Austria, and in 1999, upon approval of Parliament, the President of the Republic of Austria awarded him the title of University Professor of Austria. In 2003 he was awarded the title of Ehrenbürger (Honorary Fellow) of the Vienna University of Technology, and in 2004 was awarded the Austrian Cross of Honor for Arts and Science, First Class from the President of the Republic of Austria. In 2005 he was inducted into the Hunter College Alumni Hall of Fame, and in 2006 he was awarded the prestigious Townsend Harris Medal by the City College Alumni

Association. He was inducted into the New York State Mathematics Educator's Hall of Fame in 2009, in 2010 he was awarded the coveted Christian-Peter-Beuth Prize in Berlin, and in 2017 he received the Summa Cum Laude nemine discrepante Award from Fundacion Sebastian, A.C. in Mexico City.

He has taken on numerous important leadership positions in mathematics education locally. He was a member of the New York State Education Commissioner's Blue-Ribbon Panel on the Math-A Regents Exams, and the Commissioner's Mathematics Standards Committee, which redefined the Standards for New York State, and he also served on the New York City schools' Chancellor's Math Advisory Panel. Dr. Posamentier is a leading commentator on educational issues and continues his long-time passion of seeking ways to make mathematics interesting to both teachers, students and the general public — as can be seen from some of his more recent books:

- The Secret Lives of Numbers: Numerals and Their Peculiarities, Prometheus Books, 2022.
- Geometry in Our Three — Dimensional world (World Scientific Publishing, 2021).
- Creative Secondary School Mathematics: 125 Enrichment Units for Grades 7 to 12 (World Scientific Publishing, 2021).
- Innovative Teaching: Best Practices from Business and Beyond for Mathematics Teachers (World Scientific Publishing, 2021).
- Math Tricks: The Surprising Wonders of Shapes and Numbers (Prometheus Books, 2021).
- The Joy of Geometry (Prometheus Books, 2021).
- Mathematics Entertainments for the Millions (Prometheus Books, 2020).
- Understanding Mathematics Through Problem Solving (World Scientific Publishing, 2020).
- Math Makers: The Lives and Works of 50 Famous Mathematicians (Prometheus, 2019)
- 2. The Mathematics Coach Handbook (World Scientific, 2019)
- The Mathematics of Everyday Life (Prometheus, 2018)

- The Joy of Mathematics: Marvels, Novelties, And Neglected Gems That Are Rarely Taught in Math Class (Prometheus, 2017)
- Strategy Games to Enhance Problem-Solving Ability in Mathematics (World Scientific, 2017)
- The Circle: A Mathematical Exploration Beyond the Line (Prometheus, 2016)
- Problem-Solving Strategies in Mathematics: From Common Approaches to Exemplary Strategies (World Scientific, 2015)
- Effective Techniques to Motivate Mathematics Instruction (Routledge, 2016)
- Numbers: Their Tales, Types and Treasures (Prometheus, 2015)
- Teaching Secondary Mathematics: Techniques and Enrichment Units, 9th Ed. (Pearson, 2015)
- Mathematical Curiosities: A Treasure Trove of Unexpected Entertainments (Prometheus, 2014)
- Geometry: Its Elements and Structure (Dover, 2014)
- Magnificent Mistakes in Mathematics (Prometheus Books, 2013)
- 100 Commonly Asked Questions in Math Class: Answers that Promote Mathematical Understanding, Grades 6-12 (Corwin, 2013)
- What Successful Math Teachers Do: Grades 6-12 (Corwin 2006, 2013)
- The Secrets of Triangles: A Mathematical Journey (Prometheus Books, 2012)
- The Glorious Golden Ratio (Prometheus Books, 2012)
- The Art of Motivating Students for Mathematics Instruction (McGraw-Hill, 2011)
- The Pythagorean Theorem: Its Power and Glory (Prometheus, 2010)
- Mathematical Amazements and Surprises: Fascinating Figures and Noteworthy Numbers (Prometheus, 2009)
- Problem Solving in Mathematics: Grades 3-6: Powerful Strategies to Deepen Understanding (Corwin, 2009)
- Problem-Solving Strategies for Efficient and Elegant Solutions, Grades 6-12 (Corwin, 2008)
- The Fabulous Fibonacci Numbers (Prometheus Books, 2007)

- Progress in Mathematics K-9 textbook series (Sadlier-Oxford, 2006–2009)
- What successful Math Teacher Do: Grades K-5 (Corwin 2007)
- Exemplary Practices for Secondary Math Teachers (ASCD, 2007)
- 101+ Great Ideas to Introduce Key Concepts in Mathematics (Corwin, 2006)
- π, A Biography of the World's Most Mysterious Number (Prometheus Books, 2004)
- Math Wonders: To Inspire Teachers and Students (ASCD, 2003)
- Math Charmers: Tantalizing Tidbits for the Mind (Prometheus Books, 2003)

 **Arthur Kramer** began his primary academic life in Brooklyn, New York, and then continued his education in Manhattan at Stuyvesant High School, then earning a BME degree from Cooper Union Engineering College, followed by an MA in Mathematics Education from Columbia University and then a Ph.D. in Mathematics from New York University. Having worked as an engineer for a couple of years, he then honed his teaching skills for a short time at a high school and then went on to teach Mathematics at New York University. Soon thereafter a life-long career began at New York City College of Technology of the City University of New York teaching a variety of mathematics courses, with a specialty in the history and culture of mathematics. He also taught courses in the following departments: Physics, Electrical Engineering and Computer Engineering. During this time, he also spent several years as an adjunct Professor of Mathematics at the New School for Social Research.

While early on he was interested in mathematical research and published several articles in mathematical journals, he found his real calling in mathematics education resulting in educational grants and programs at New York City College of Technology. In particular, he developed a curriculum for a program called "Access for Women,"

which held weekend sessions for high school girls where they did basic experiments in various engineering disciplines hoping to motivate them to enter one of these fields.

Professor Kramer has published five textbooks in technical mathematics, his latest being "Mathematics for Electricity and Electronics" and continues to teach as an adjunct professor in the Electrical Engineering department at NYC College of Technology

When not devoting time to education and writing, he is an accomplished offshore sailor who has sailed his Ketch "Sweet Harmony" along most of the east coast and Nova Scotia, to Bermuda several times, to the Caribbean, across the Atlantic and throughout the Mediterranean. He enjoys ballroom dancing and has recently discovered a pleasurable pastime playing tennis.

Contents

Introduction

Whenever the topic of mathematics is mentioned, people tend to indicate their weakness in the subject as a result of not having enjoyed its instruction during their school experience. Many students unfortunately do not have very positive experiences when learning mathematics, which can result from teachers who have a tendency "to teach to the test". This is truly unfortunate for several reasons. First, basic algebra and geometry, which are taken by almost all students, are not difficult subjects, and all students should be able to master them with the proper motivational instruction. Second, we live in a technical age, and being comfortable with basic mathematics can certainly help you deal with life's daily challenges. Other, less tangible reasons, are the pleasure one can experience from understanding the many intricacies of mathematics and its relation to the real world, experiencing the satisfaction of solving a mathematical problem and discovering the intrinsic beauty and historical development of many mathematical expressions and relationships. These are some of the experiences that this book is designed to deliver to the reader.

The book offers 101 mathematical gems, some of which may require a modicum of high school mathematics and others, just a desire to carefully apply oneself to the ideas. Many folks have spent years encountering mathematical terms, symbols, relationships and other esoteric expressions. Their origins and their meanings may never have been revealed, such as the symbols $+$, $-$, $=$, π. ∞, $\sqrt{\ }$, Σ, and many others. This book provides a delightful insight into the origin of mathematical symbols and popular theorems such as the Pythagorean Theorem and the Fibonacci sequence, common mathematical mistakes and curiosities, intriguing number relationships and some of the different mathematical procedures in various countries. The book uses a historical and cultural approach to the topics, which enhances the subject matter and greatly adds to its appeal. The mathematical material can, therefore, be more fully appreciated and understood by anyone who has a curiosity and interest in mathematics, especially if in their past experience they were expected to simply accept ideas and concepts without a clear understanding of their origins and meaning. It is hoped that this will cast a new and positive picture of mathematics and provide a more favorable impression of this most important subject resulting in a different experience than what many may have previously encountered. It is also our wish that some of the fascination and beauty of mathematics shines through in these presentations.

CHAPTER 1

All About Numbers and Symbols

UNIT 1

Some Ancient Number Systems

Our decimal system is based on the number ten only because we have ten fingers (digits) and they were, and still can be, used for counting. But has that always been the number base used by past civilizations? Though the numerals 1 to 10 played an important part in the number systems of most civilizations, the ancient Sumerians and Babylonians based their number system on 60, the Aztecs and the Mayans on 20, and some early civilizations used a quinary system based on 5 using one hand for counting. The binary system based on 2, the octal system based on 8 and the hexadecimal system based on 16 have also found important applications with the advent of computers. Number systems based on 5, 10 and 20 all stem from our human digits but there is nothing special about these number bases. If we possessed 6 digits on each hand, as some polydactyl cats (Figure 1.1) who have 6 claws on each leg, 12 would serve as a perfectly fine number base. It would even be easier to do calculations with fractions, since 12 has divisors of 2, 3, 4 and 6, which is twice as many divisors as 10 has.

The Sumerians in Mesopotamia, which is now Iraq, previously used a base 60 system. They began using clay cones as early as the 4th millennium BCE (Before the Common Era) to record their

Figure 1.1   Polydactyl kitten

Figure 1.2   Sumerian mathematics

mathematics, as shown in Figure 1.2. They most likely invented the cuneiform symbols as shown in Figure 1.2 and Figure 1.3.

The Babylonians improved on the base 60 system used by the Sumerians by employing a positional-number system where larger values were placed on the left, as we do in our decimal system. This was a great achievement by the Babylonians as it was a more advanced number system than the ones used by the Egyptians, Greeks and

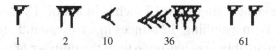

Figure 1.3   Babylonian number symbols

Romans, who did not use place-values to represent their numbers. What is also very interesting here is that the Babylonians used only two symbols to represent their numbers as opposed to our current system which uses 10 symbols.

The place values from right to left were 1, 60, 60^2, 60^3, etc. There was no zero in the system, but a space was used to represent a zero. For example, the numbers 2 and 61 were both written with two characters each representing one unit, but 61 is written with a space between the characters. Figure 1.3 shows the two symbols, which represent one and ten, and how the numbers 2, 36 and 61 appear in their system.

A problem does arise, however, with the number 60 and the number 2, since both are written with a single character representing one unit and there is no zero to fill the first space for the number 60. We can only assume that since it does not occur very often, only for numbers divisible by 60, somehow the Babylonians overcame this problem. However, there is evidence that later Babylonian civilizations did invent a symbol to indicate an empty place so the lack of a zero could not have been totally satisfactory to them. Today we still have the vestiges of the sexagesimal system (base 60) in our measurement of time and angles.

The question that remains is, why was a sexagesimal base system first used by the Sumerians, and then the Babylonians, who replaced their civilization? Many theories have been advanced, though we have no definitive proof. However, here are some of the noteworthy speculations. Theon of Alexandria (ca. 335–405) proposed the idea that 60 is the smallest number that has the divisors 1, 2, 3, 4, 5, and 6. It could then serve to make working with fractions easier than a base 10 system. On this note, 12 has the divisors 1, 2, 3, 4, and 6 and could also be a possible number base. Although no civilization has done that, 12 appears often in the UK and US systems involving measurement, money and weights. There are 12 inches in a foot, 12 hours in a half day, 12 pennies in a shilling, 12 ounces in a troy pound etc.

Another plausible idea has to do with the number of days in a year that might have been thought to be 360. However, we must give the ancients more credit than to think that they were not aware that the year was longer than 360 days. Here we quote from an article by J. J. O'Connor and E. F. Robertson:

> *"I just do not believe that anyone ever chose a number base for any civilization. Can you imagine the Sumerians setting up a committee to decide on their number base — no, things just did not happen in that way. The reason has to involve the way that counting arose in the Sumerian civilization, just as 10 became a base in other civilizations who began counting on their fingers, and twenty became a base for those who counted on both their fingers and toes. Here is one way that it could have happened. One can count up to 60 using your two hands. On your left hand there are three parts on each of four fingers (excluding the thumb). The parts are divided from each other by the joints in the fingers. Now one can count up to 60 by pointing at one of the twelve parts of the fingers of the left hand with one of the five fingers of the right hand. This gives a way of finger counting up to 60 rather than to 10. Anyone convinced?"*[1]

The Egyptians, Greeks and Romans all used number systems based on 10. The Egyptian civilization, a contemporary of the Babylonian, used picturesque hieroglyphic numbers for one of their number systems. A different symbol represented each power of 10 and numbers were written from left to right as shown in Figure 1.4.

The symbol for 1 is a vertical staff. A lotus plant is the symbol representing 1000, a pointing finger represented 10,000, a frog represented 100,000, and a man in astonishment represented 1,000,000. The Egyptians applied an additive concept to represent numbers. No more than 4 symbols were written on one line so the numbers 4 through 9 have the symbols written above each other.

---

[1] https://mathshistory.st-andrews.ac.uk/HistTopics/Babylonian_numerals/

200,310

Figure 1.4 Egyptian Hieroglyphic numbers

α	β	γ	δ	ε	F	ζ	η	θ
1	2	3	4	5	6	7	8	9

ι	κ	λ	μ	ν	ξ	ο	π	Ϙ
10	20	30	40	50	60	70	80	90

ρ	σ	τ	υ	φ	χ	ψ	ω	ϡ
100	200	300	400	500	600	700	800	900

Figure 1.5 Ancient Greek numbers

The Greeks learned from the mathematics of the Babylonian and Egyptian civilizations but developed a much more introspective approach toward the subject. They no longer were content to use numbers merely for calculations. It became important to study relationships, to question results, and try to prove them. Their great accomplishments were not made any easier with their number system which was based on their alphabet and was more difficult to work with than the Babylonian system. The 24 Greek letters and some other symbols represented their numbers as shown in Figure 1.5. The Hebrews and the Syrians also used their alphabet for representing numbers, but the Greeks may have been the first to do so. Why they departed from the symbols used by the Babylonians and Egyptians is paradoxical. One advantage is that in the Greek system, numbers could be written more compactly. For example, 43,678 was written: δM, γχοη where M was used for 10,000. Because of their great contributions in mathematics, the Greek number symbols and their defects existed for some time, even after better systems came into being. Many years later in the early part of the 14th century, a Calabrian monk still wrote arithmetical proofs of some of the material in Euclid's Elements using the Greek language and their number system.

The great Roman empire for some reason never adopted the alphabetic numeral systems that were used by the Syrians, Hebrews and the Greeks. The origin of their symbols remains somewhat speculative. Around the 4th century BCE, their number system developed, and it is thought to have originated from Etruscan numbers and the tally sticks used by shepherds to mark the size of their flock. Their numbers, however, were in use by Europeans for almost two thousand years, longer than our present Hindu-Arabic system has been in existence.

In Roman times, the concept of zero did not exist as a number, consequently, the numeral placement employed both subtraction and addition. The largest Roman numeral is **M** for 1000 and they used no more than three of the same letters consecutively in a number. The largest Roman number then turns out to be **MMMCMXCIX** or 3999, where C = 100, X = 10 and I = 1. However, it developed later that using a line above a symbol multiplies it by 1,000, so that 1,000,000 could be written as $\overline{M}$.

One can still find many appearances of Roman numerals today such as on the pillars of docks, on the hulls of ships marking the water level (Figure 1.6), on films indicating the year of production, in the military for designating units, marking quadrants on a graph to distinguish from the Arabic numerals, and their ubiquitous presence on the cornerstones of buildings. It was not difficult to do addition and subtraction in the Roman system, but multiplication, division and other arithmetic operations presented more challenges. Hence, the Roman system did not lead to many advances in mathematics. It was the Hindu-Arabic system around 600 CE which gave birth to our modern algebra.

Why is there still the continued practice of using Roman numerals on buildings? One explanation goes back to the birth of our nation.

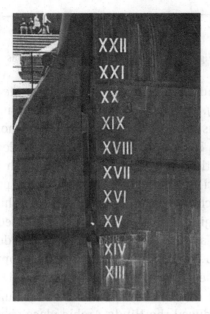

Figure 1.6   Roman numerals on stern of the ship *Cutty Sark* showing the draft in feet. The numbers range from 13 to 22, from bottom to top

Our Democracy embraced classic Greek and Roman ideals and our early architecture emulated their style in our buildings.

UNIT 2

The Origin of Our Numerals

We tend to take for granted the number system we use, but it is worth taking the time to appreciate where it originated and how it got to dominate our society. The place-value number system that we use today has a rather interesting beginning. We call it the Hindu-Arabic number system because it was devised by Indian mathematicians during the first four centuries of the common era (CE) and then later adopted by Arabic mathematicians in the 9th century. It was popularized by two Arabic books: Muḥammad ibn Mūsā al-Khwārizmī (780–850) wrote *The Compendious Book On the Calculation with Hindu Numerals* (ca. 825) and the other book by Al-Kindi (801–873) who

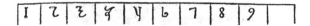

Figure 1.7 Early Hindu-Arabic numerals

popularized the Indian numerals in his book *On the Use of the Hindu Numerals* (ca. 830). The early numerals are shown in Figure 1.7.

Leonardo of Pisa (ca. 1170–ca. 1245), better known as Fibonacci, spent his youth in Bugia, a town on the Barbary Coast of Africa and traveled through the Arab world interacting with mathematicians. Although Fibonacci wrote several books, the one of particular interest is *Liber abaci,* which was first published in 1202. This extensive book is full of very engaging problems. The book was based on the arithmetic and algebra that Fibonacci had accumulated during his travels largely in the Arab world. The book was widely copied and imitated and although it is best known for having introduced the famous series of numbers today referred to as the Fibonacci numbers (see page 333), it also introduced the Hindu-Arabic place-valued decimal system to the Western world, essentially Europe. The book was widely used for the better part of the next 2 centuries — a best seller!

Fibonacci begins *Liber abaci* with the following:

"The nine Indian[2] figures are:

$$9\ 8\ 7\ 6\ 5\ 4\ 3\ 2\ 1.^{3}$$

With these nine figures, and with the sign 0, which the Arabs call zephyr, any number whatsoever is written, as demonstrated below. A number is a sum of units, and through the addition of them the number increase by steps without end. First one composes those numbers, which are from one to ten. Second, from the tens are made those numbers, which are from ten up to one hundred. Third, from the hundreds are made those numbers, which are from one hundred up to one thousand. ... and, thus, by an unending sequence of steps, any number whatsoever is constructed by joining the pre-

---

[2] Fibonacci used the term Indian figures for the Hindu numerals.
[3] It is assumed that Fibonacci wrote the numerals in order from right to left, since he took them from the Arabs who write in this direction.

ceding numbers. The first place in the writing of the numbers is at the right. The second follows the first to the left."

Despite their relative facility, these numerals were not widely accepted by merchants who were suspicious of those who knew how to use them. They were simply afraid of being cheated. We can safely say that it took over 50 years for these numerals to begin to replace the Roman numerals in Europe.

There are many other aspects of our current usage of mathematics that had its beginnings in *Liber abaci*. For example, simultaneous linear equations were also presented in this book. Many of the problems that Fibonacci considers in *Liber abaci* were similar to those appearing in Arab sources. This does not detract from the value of the book, since the solutions to these problems really carry the main message. As a matter of fact, a few mathematical terms — common in today's usage — were first introduced in *Liber abaci*. Fibonacci referred to "factus ex multiplicatione" and from this first sighting of the word, we speak of the "factors of a number" or the "factors of a multiplication". Another example of words whose introduction into the current mathematics vocabulary seems to stem from this famous book are the words "numerator" and "denominator."

UNIT 3

The History of Zero

The mysterious symbol zero and its concept is a relative newcomer in the history of mathematics and did not find its way into the mainstream of western mathematics until the Middle Ages. Beginning its life as a placeholder and evolving through many centuries and diverse cultures, our computerized world could not exist without it. The idea of a mathematical symbol representing "nothing" was not easy for ancient civilizations to accept, as early number systems were used primarily for practical applications. Some of these early number systems simply used a space to represent no-place-value as in the base 60 Babylonian system. For example, Figure 1.8 (left) shows the number $3601 = (1)60^2 + (0)60 + 1$ in the early Babylonian system. Each

𒁹 𒁹   𒁹 " 𒁹

Figure 1.8   Early and late Babylonian number $3601 = 60^2 + 0 + 1$

symbol represents a place value of 1 separated by a space. Sometime around 400 BCE, the Babylonians began to use 2 wedge symbols to represent zero as a placeholder so 3601 would then be written as shown in Figure 1.8 (right).

The abstract Babylonian way of writing numerals had a profound influence on the scholars of antiquity. Greek astronomers, although used to a decimal system, translated the cuneiform script into their own "alphabetical" way of writing digits. However, they adopted the Babylonian system, for expressing fractional parts of 60. It would have been too much work to convert thousands of ancient astronomical tables into a decimal system, which is the reason that we still measure units of time, as well as units of angle size, in a sexagesimal system, dividing hours and degrees into 60 minutes and 60 seconds.

Zero serves two roles in our number system. One as a placeholder, indicating no-value in a unit place, as in the number 2021. The other, as a real number in the decimal system. The concept of zero as a placeholder was clearly understood by ancient civilizations, as early uses of mathematics were in solving concrete problems such as land area, construction, possessions, etc. Its acceptance as a real number is a more abstract idea and came much later. As late as the 7th century, we first find the use of zero, as we know it today, appearing in Indian mathematics. However, it is fair to say that the Indian system did owe many of its ideas to earlier systems. Indian mathematicians used the concept of zero as both a placeholder and a real number. During the 7th century, the Indian mathematician Brahmagupta (598–668), developed rules for zero and negative numbers such as:

The sum of two positives is positive, the sum of two negatives is negative; of a positive and a negative, the sum is their difference; if they are equal, it is zero.

Curiously, Brahmagupta also stated that *zero divided by zero is zero*, though he gives no explanation how he arrived at that, and today we consider it an undefined quantity.

One of the first genuine written records of zero resembling the symbol in use today, is found inscribed on a stone tablet dated 876 CE and relates to a garden in the town of Gwalior, 400 km south of Delhi, India. The garden produced enough flowers to give 50 garlands per day to the local temple and the number 50 differs little from our present number 50. These contributions provided the foundation for present-day algebra as they spread west to the Islamic world and east to the Orient. In China, the mathematician Qin Jiushao (1202–1261) wrote a *Mathematical Treatise in Nine Sections* in which can be found the symbol 0 for zero. In Italy, Leonardo of Pisa (ca. 1170–1250), better known by the name Fibonacci, and one of the most brilliant mathematicians of the Middle Ages, introduced the symbol for zero, 0, in his treatise *Liber Abaci* in 1202. The Hindu-Arabic symbols 1 to 9 he introduced as numbers but was not confident to assign to zero the status of a number and referred to it merely as a "sign." The sources for *Liber Abaci* were gathered principally from the Arabic world, which he visited on many journeys, but he modified and enlarged the material he obtained through his own genius. *Liber Abaci* was the first Western book to employ the Hindu-Arabic numerals and demonstrated the superiority of the system to both merchants and mathematicians (see Figure 1.9).

The use of Hindu-Arabic numerals took time to be universally accepted in western culture. Mathematicians were the first to accept zero and the Hindu-Arabic numerals while merchants continued to use Roman numerals up to the 16th century. Eventually the Hindu-Arabic numerals became more common in Europe. At this point, it is worth quoting the French mathematician Pierre Simon Laplace's (1749–1827) praise for our number system:

> *The ingenious method of expressing every possible number using a set of ten symbols (each symbol having a place value and an absolute value) emerged in India. The idea seems so simple nowadays that its significance and profound importance is no longer appreciated. Its simplicity lies in the way it facilitated calculation and placed arithmetic foremost amongst useful inventions. The importance of this invention is more readily appreciated when one considers that it was beyond the two greatest men of antiquity, Archimedes and Apollonius.*[4]

---

[4]An Overview of Indian Mathematics, MacTutor History of Mathematics Archive, School of Mathematics and Statistics, University of St Andrews, Scotland.

Figure 1.9   A page from Fibonacci's *Liber Abaci* showing calculations in the margin

One can only imagine what arithmetic would be like today without the ubiquitous symbol for "nothing" — the zero.

UNIT 4

The Origin of the Equal Sign

Perhaps the most frequently used mathematical symbol is the equal sign (=). Its first appearance was in an English text by the Welsh mathematician Robert Recorde (1512–1558). The page in the book *The Whetstone of Witte* (London, 1557), where it appears is shown in Figure 1.10. The last paragraph on the page introduces the equal sign and the first line below it indicates the following: $14 \times 4 + 15 = 71$, which is where the = is used for the first time.

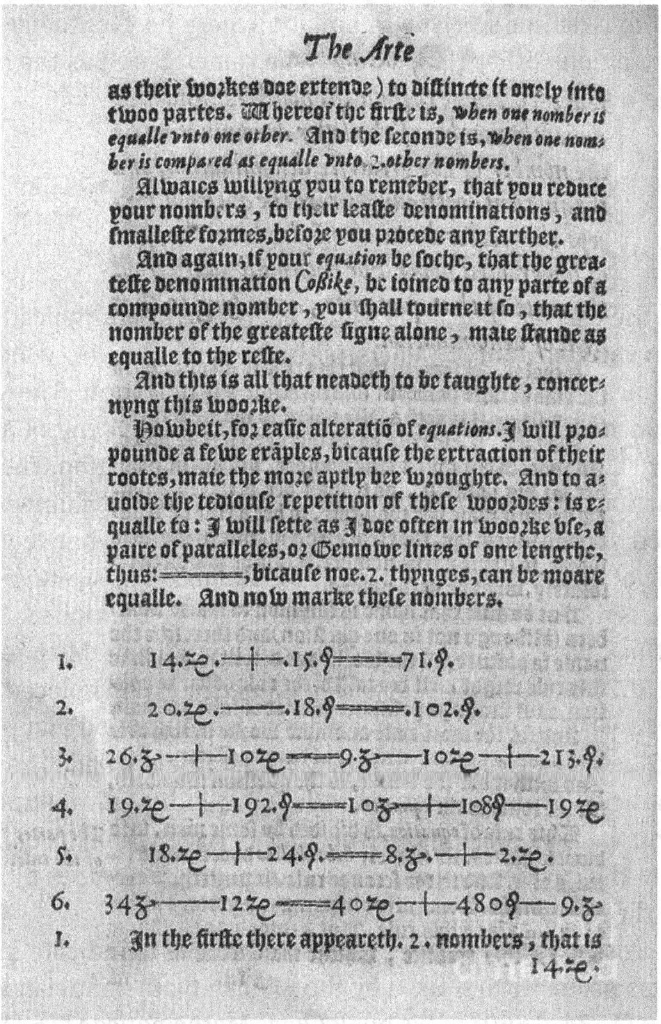

The Arte

as their workes doe extende) to distincte it onely into twoo partes. Whereof the firste is, *when one nomber is equalle vnto one other.* And the seconde is, *when one nomber is compared as equalle vnto .2. other nombers.*

Alwaies willyng you to remeber, that you reduce your nombers, to their leaste denominations, and smalleste formes, before you procede any farther.

And again, if your *equation* be soche, that the greateste denomination *Cosike*, be ioined to any parte of a compounde nomber, you shall tourne it so, that the nomber of the greateste signe alone, maie stande as equalle to the reste.

And this is all that neadeth to be taughte, concernyng this woorke.

Howbeit, for easie alteratiō of *equations*. I will propounde a fewe exāples, bicause the extraction of their rootes, maie the more aptly bee wroughte. And to auoide the tediouse repetition of these woordes: is equalle to: I will sette as I doe often in woorke use, a paire of paralleles, or Gemowe lines of one lengthe, thus: =====, bicause noe .2. thynges, can be moare equalle. And now marke these nombers.

1. 14.ze.—+—.15.ϕ=====71.ϕ.
2. 20.ze.——.18.ϕ=====.102.ϕ.
3. 26.ʒ.—+—10ze=====9.ʒ.——10ze—+—213.ϕ.
4. 19.ze—+—192.ϕ=====10ʒ—+—108ϕ——19ze
5. 18.ze—+—24.ϕ.=====8.ʒ.—+—2.ze.
6. 34ʒ——12ze=====40ze—+—480ϕ——9.ʒ.

1. In the firste there appeareth. 2. nombers, that is 14.ze.

Figure 1.10

There you will also see the plus (+) and minus sign (–) for the first time in an English book. Although he was trained as a physician at Cambridge University, Recorde eventually immersed himself in mathematics, which he taught at Oxford University. He had experienced an influential life, which included being controller of the Bristol Royal Mint. His life unfortunately came to a tragic end when he insulted the Earl William Herbert of Pembroke, whom he accused of malfeasance. The court found this accusation so insulting that he was consequently confined to a debtor's prison in London where he eventually died in 1558. Yet, among other things he will be remembered as the originator of the = sign.

UNIT 5

The Birth of Multiplication and Its Symbol

Perhaps the earliest indication of the operation of multiplication was found on the Babylonian tablets with the symbol *A-DU*, which indicated "times" or as we know it today: multiplication. The ancient Greeks, such as Diophantus (208–292 CE), used no symbol and just wrote two numbers side-by-side to indicate multiplication. The Indian mathematician Bhaskara (1114–1185) often wrote the numbers to be multiplied side-by-side and on occasion placed a dot between them. Leonardo of Pisa, better known today as Fibonacci (1170–1275), often used a straight line to indicate two numbers to be multiplied.

As time went on, the German mathematician Michael Stifel (1487–1567) in his book *Deutsche Arithmetica* (1545) placed a capital *M* between two numbers that were to be multiplied and placed a capital *D* between two numbers that were to be divided. The French mathematician François Viete (1540–1603) indicated multiplication of two numbers by placing the word *"in"* between them as in *"A in B."* Gradually, the custom evolved when multiplying a number by a letter to write them together without a space, such as $7x$.

The cross (×) that we are accustomed to use to indicate multiplication was probably first used by the English mathematician William Oughtred (1574–1660) in his book *Clavis Mathematicae* (1631). He is also very well known for his having developed the first slide rule, a calculating device that preceded the modern calculator. Today we use a variety of symbols to indicate multiplication. When a number and a letter, or two letters are written beside one another it indicates multiplication. Two adjacent parenthetical expressions also indicate multiplication as does a dot between two numbers. However, the classical multiplication symbol still continues to be the cross $\times$.

The German mathematician Gottfried Wilhelm Leibniz (1646–1716) introduced the dot instead of the symbol $\times$ and wrote to the Swiss mathematician Johann Bernoulli (1667–1748) on July 29, 1698 "I do not like the $\times$ as a symbol for multiplication, as it is easily confused with the letter x. Often, I simply relate two quantities to be multiplied with an interposed dot and indicate multiplication by $AB \cdot CD$. Hence, in designating ratio, I use 1 point not 2 points, which I use at the same time for division." This is probably not the first time that a dot was used for multiplication, since it was already used in 1631 by the English mathematician Thomas Harriot (1560–1621) in his *Artis Analyticae Praxis.* However, we are not sure if Harriot meant for the dot to represent multiplication.

Now that we have shown how the symbols used for multiplication have evolved, let's consider an old algorithm for multiplication. One of the earliest algorithms[5] for doing multiplication appears in the *Rhind Mathematical Papyrus* (ca. 1650 BCE), where the Egyptians recorded methods of calculating quantities. This method, which we will again consider as it was used by the Romans to do multiplication (see "How the Romans did multiplication" page 109), is still probably used today in parts of Russia and is known as *The Russian Peasant's Method of Multiplication.* It demonstrates a rather strange and primitive algorithm for doing multiplication. Yet, it is interesting to see a very different style from that which we are accustomed to using today. It is actually simple, yet somewhat cumbersome. Consider the problem of finding the product of 43×92.

---

[5]We define an algorithm as a step-by-step problem-solving procedure, especially an established, recursive computational procedure for solving a problem in a finite number of steps.

We begin by setting up a chart of 2 columns with the 2 numbers of the product in the first row. In Figure 1.11, we have the 43 and 92 heading up the columns. One column will be formed by doubling each number to get the next, while the other column will take half the number and drop the remainder. For convenience, our first column will be the doubling column, the second column will be the halving column. Notice that by halving the odd number such as 23 (the third number in the second column) we get 11 with a remainder of 1, and we simply drop the 1. The rest of this halving process should now be clear.

We now identify the odd numbers in the halving column (here the right column), then get the sum of the partner numbers in the doubling column (in this case the left column). These are highlighted in bold type. This sum gives you the originally required product of 43 × 92. In other words, with the Russian Peasant's Method we get 43 × 92 = 172 + 344 + 688 + 2752 = 3956.

In the example above, we chose to use the first column, as the doubling column and the second column as the halving column. We could also have done this Russian Peasant's Method by halving the numbers in the first column and doubling those in the second (see Figure 1.12).

To complete the multiplication, we find the odd numbers in the halving column (in bold type), and then get the sum of their partner

43	92
86	46
172	**23**
344	**11**
688	**5**
1376	2
2752	**1**

Figure 1.11

43	92
21	184
10	368
5	736
2	1472
1	2944

Figure 1.12

numbers in the second column (now the doubling column). This gives us $43 \times 92 = 92 + 184 + 736 + 2944 = 3956$.

Clearly, we are not expected to do multiplication in this high-tech era by using the Russian Peasant's Method. However, it should be fun to observe how this primitive system of arithmetic actually work. Explorations of this kind are not only instructive but should be entertaining.

In the following, you see what was done in the above multiplication algorithm.

$$
\begin{aligned}
*43 \times 92 &= (21 \times 2 + 1)(92) &= 21 \times 184 &+ 92 = 3956 \\
*21 \times 184 &= (10 \times 2 + 1)(184) &= 10 \times 368 &+ 184 = 3864 \\
10 \times 368 &= (5 \times 2 + 0)(368) &= 5 \times 736 &+ 0 = 3680 \\
*5 \times 736 &= (2 \times 2 + 1)(736) &= 2 \times 1472 &+ 736 = 3680 \\
2 \times 1472 &= (1 \times 2 + 0)(1472) &= 1 \times 2944 &+ 0 = 2944 \\
*1 \times 2944 &= (0 \times 2 + 1)(2944) &= 0 &+ \underline{2944} = 2944 \\
& & & \quad\quad 3956
\end{aligned}
$$

For those familiar with the binary system (base 2), one can also explain this Russian Peasant's Method with the following representation.

$$(43)(92) = (1 \times 2^5 + 0 \times 2^4 + 1 \times 2^3 + 0 \times 2^2 + 1 \times 2^1 + 1 \times 2^0)(92)$$
$$= (92)(2^5) + (92)(2^3) + (92)(2^1) + (92)(2^0)$$
$$= 2944 + 736 + 184 + 92$$
$$= 3956$$

Whether or not you have a full understanding of the discussion of the Russian Peasant's Method of multiplication, you should at least now have a deeper appreciation for the multiplication algorithm you learned in school. There are many other multiplication algorithms, yet the one shown here is perhaps one of the strangest and it is through this strangeness that we can appreciate the powerful consistency of mathematics that allows us to conjure up such an algorithm.

UNIT 6

The Origin of the Division Symbols

It is not clear exactly where our division symbols originated. The ancient Babylonians used words rather than symbols to express division. Similarly, the Greeks also had no sign to indicate division. For example, Diophantus (ca. 208–292 CE) separated the dividend from the divisor using the word *uoplov,* which was to mean "divided by." In 1881, an ancient Indian manuscript written on birch bark was found in the city of Bakhshali (today located in Pakistan) and dates back to the 1st century of the common era (see Figure 1.13). This is perhaps the oldest document that exhibits numerals for place value as well as a symbol for zero. Here division is indicated with the abbreviation "*bha*" to represent "*bhaga,*" meaning apart.

Later, the Hindus simply wrote the divisor below the dividend to indicate division. Still no symbol has been used to this point to indicate division. It was perhaps the 12th century Muslim mathematician from Morocco, Abu Bakr al-Hassar, who was the first to use the horizontal line to indicate division. However, we must admit, his usage was rather strange. For example, he would indicate division as follows "Write the denominators below a horizontal line and over each of

Figure 1.13

them the parts belonging to it; for example, if you are told to write 3 aspects and $\frac{1}{3}$ of $\frac{1}{5}$ it would be written as follows: $\frac{3}{5}\frac{1}{3}$."

In his 1202 book *Liber abaci*, Leonardo of Pisa, the famous mathematician better known as Fibonacci (1170–1240) used the horizontal fraction bar to indicate division. He defined his system as follows: "When above a number a line is drawn and above that another number is written then the above number indicates the parts of the below number." These were respectively the numerator and denominator as we know them today. Fibonacci then gives the example $\frac{1}{2}\frac{5}{6}\frac{7}{10}$, which is to mean seven-tenths, and five-sixths of one-tenth, and one-half of one-sixth of one-tenth. As odd as this appears, we must bear in mind that the ancient Egyptians only worked with unit fractions and the fraction $\frac{2}{3}$. Up to this period, division was well defined as we know it today.

The typical division symbol, $7\overline{)98}$ still used in the United States today was first introduced by the German mathematician Michael Stifel (1487–1567) in his 1544 book *Arithmetic Integra*. The English mathematician William Oughtred (1574–1660) in his book *Clavuis mathematicae* also used parentheses in the following way: $\frac{4}{3}\big)\frac{2}{3}\big(\frac{2}{1}$, which indicates $\frac{4}{3}$ divided by $\frac{2}{3}$ equals 2. In his 1845 book *Deutsche arithmetica,* Stifel used the Gothic letters 𝕸 and 𝕯 to indicate multiplication and division. The French mathematician J. E. Gallimard (1685–1771) used an inverted D to indicate division as in the example: $45 \ \Box \ 5 = 9$.

One of the common signs used for division in the United States is the symbol ÷, which was first introduced in 1659 by the Swiss mathematician Johan Heinrich Rahm (1662–1676) in his book *Teutsch Algebra.* Curiously, the book was translated into English and even though it was praised by the German mathematician Wilhelm Leibniz (1646–1716), it did not enjoy much popularity in Switzerland, but flourished in England and, thus, the symbol became very popular then in the United States. However, it still did not achieve popularity on the European continent. We are still using that symbol today — as most people will recognize it — to indicate the division key on a calculator.

As much as Leibniz was in favor of the division symbol ÷, in 1684 he introduced for the first time the double dot (:), as a division symbol in an article entitled *Acta eruditorum.* This symbol is today used as a colon. In 1710, the *Miscellanea Berolinensia*, a compendium of articles by the Royal Society of Science in Berlin featured an article by Leibniz where he writes "according to common practice, the division is sometimes indicated by writing the divisor beneath the dividend, with a line between them; thus, *a* divided by *b* is commonly indicated by $\frac{a}{b}$. Very often however, it is desirable to avoid this and to continue on the same line, but with the interposition of 2 points; so that *a:b* means *a* divided by *b*." Most of continental Europe now uses the colon: to represent division as well as ratio. In Latin America a backward slash was used to indicate division as shown here: $\left(\frac{8}{7} \backslash 2\right) = \frac{8:2}{7} = \frac{4}{7}$.

When we thought everything now was as clear as possible, the French mathematician François Peyrard (1760–1822), who helped reform the French education system, developed another symbol for division in his book *Arithmetique de Bezout*, where the example 14464 divided by 8 is written as follows: $14464 \left| \frac{8}{1808} \right.$.

In 1845, Augustus De Morgan (1806–1871) was the first to use the slash to indicate division, such as 8/2, which is still today useful, since it does not require more vertical space as does the horizontal-bar fraction. Thus, there exists a broad history of how division was noted in the past, and the long road that it took to get to where we are today with our modern symbols for division.

UNIT 7

The Origins of Fractions

Working with fractions is an elusive skill that some do not master in elementary school and continue to be plagued with them in high school, and for some even throughout college. The ancients also struggled with fractions but could not ignore their need and application. Maybe this difficulty can be found in the word fraction which comes from the Latin "fractio," which means to break. The earliest use of fractions is found in the *Ahmes* or *Rhind Papyrus* written in Egypt around 1550 BCE and named after Alexander Henry Rhind (1833–1863), a Scottish Egyptologist who purchased the papyrus in 1858 in Luxor, Egypt. Ahmes was only the scribe, and it is believed that the material dates back to about 2000 BCE. The papyrus contains considerable information about ancient Egypt and is our major source of knowledge about Egyptian mathematics, which like our present number system, was based on 10. All the fractions used in Egypt were only unit fractions having numerators of 1. The only exception was $\frac{2}{3}$. The "fraction line" is represented in hieroglyphics by an elliptical symbol ⬭, and a fraction such as $\frac{1}{4}$ looks like: ⬭. The fraction $\frac{2}{3}$ had a special symbol ⊤ and seems to be inspired by the reciprocal of $1\frac{1}{2}$: | ⬭. It is not clear why only this unique non-unit fraction was employed except maybe because of its frequent use. The Egyptians devised the intricate technique of representing non-unit fractions as a sum of unit fractions. For example, in problem 3 of the *Rhind Papyrus* $\frac{3}{5}$ is written as $\frac{1}{2}+\frac{1}{10}$.

It was the Babylonians, and their more advanced placeholder system based on 60, who introduced non-unit fractions. The fractional places of their system then had denominators of 60, 60^2, 60^3, etc. However, they did not have a symbol for the sexagesimal point (the equivalent of our decimal point), and it may have required some understanding as to the context to distinguish a whole number from a fraction. For example, consider the Babylonian numerals in Figure 1.14 consisting of unit symbols. This could represent the whole number: $2 \times 60 + 1 = 61$ or the fraction: $2 + \frac{1}{60} = \frac{121}{60}$. Somehow, the scribes were able to differentiate values. It is possible that the amount of space between the numerals was a factor that enabled them to distinguish the numbers correctly.

Figure 1.14   Babylonian numbers 61 or $\frac{121}{60}$

In some ways, their fractional system was more advanced than our decimal system. This is because a decimal fraction that has a denominator which is a power of 10, can only be represented as a finite decimal fraction when the denominator has no prime factors other than 2 or 5. On the other hand, 60 has the divisors 2, 3 and 5, which allows for more fractions in the Babylonian system to be expressed as finite sexagesimal fractions. Figure 1.15 compares unit fractions in the decimal system with unit fractions in the sexagesimal system where a comma is used for a sexagesimal point. It can be seen that there are more fractions in the Babylonian system that can be expressed as finite fractions. Note that $\frac{1}{8} = \frac{7}{60} + \frac{30}{3600}$, and $\frac{1}{9} = \frac{6}{60} + \frac{40}{3600}$.

System	1/2	1/3	1/4	1/5	1/6	1/8	1/10
Decimal	0.5		0.25	0.20		0.125	0.10
Sexagesimal	,30	,20	,15	,12	,10	,7,30	,6

Figure 1.15

Civilizations after the Babylonians did not advance the use of fractions until the Hindu-Arabic system that we use today was developed in the Middle East and found its way into Europe around the 16th century. The Greeks and the Romans had number systems less advanced than the Babylonians and more difficult to work with. The Romans used words to express fractions which were based on a denominator of 12, their unit of weight. The fraction $\frac{1}{12}$ was called "uncia" and the number 1 was equal to 12 unica. The Indian civilization significantly advanced the development of numbers, using 10 symbols including 0 during the 7th century, and with the input of the Arabs, fractions began to be written similar to the ones we commonly use now. Early Indian fractions were written with the numerator above the denominator but without a horizontal line. Thus $\frac{1}{3}$ was written as $_3^1$. It was the Arabs who added the fraction line which was drawn sometimes on a slant and sometimes horizontally. The first appearance of the fraction line is found in the writings of the Arabic author al-Hassar in the 12th century followed closely by Fibonacci in his famous book *Liber Abaci* in 1202. Fractions, as we know them today, are relatively new in mathematics, having only been around a little more than 800 years in western mathematics.

UNIT 8

The Emergence of Exponent Notation

It is believed that the idea of squaring and cubing goes back to Babylonian times based on the excavation of tablets dating back to 2300 BCE.

The Babylonian numbering system (see page 370) was based on 60 and the number 2,27 in the Babylonian system means $(2 \times 60) + 27 = 147$. Then $147^2 = 21,609$, which is equal to $(6 \times 3600) + (0 \times 60) + 9 = 21,609$. A space represented a zero for which there was no symbol (see Figure 1.16).

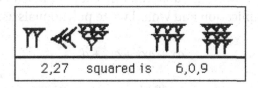

Figure 1.16

The ancient Greek mathematician Euclid (ca. 325–265 BCE) first employed the word "power" for the square erected on a line after the use by the leading mathematician and astronomer Hippocrates of Chios (470–410 BCE). During the 9th century, the Persian mathematician, Muḥammad ibn Mūsā al-Khwārizmī (780–850), used the terms *mal* for a square and *kahb* for a cube, which was shortened centuries later to *m* and *k* as shown in the work of Abū al-Ḥasan ibn Alī al-Qalaṣādī (1412–1486), a Muslim from Moorish Spain. In the late 16th century, Jost Bürgi (1552–1632), born in Lichensteig, Toggenburg, which is now part of Switzerland, used Roman numerals for exponents such as 2^V = 32. The word "exponent" was introduced by Michael Stifel (1487–1567), a German Monk, in 1544.

Around this time Robert Recorde (1512–1558), the Welsh mathematician and physician, proposed three mathematical terms by which any power, or exponent, greater than 1 could be expressed. The term, *zenzic*, which is an old German spelling of the medieval Italian word *censo*, was used to mean "squared" and the term *cubic* was used to mean the third power. The term *sursolid*, which comes from the French *sur + solid*, was used to indicate a power that is a prime number greater than 3, the smallest of which is five. Sursolids were as follows: 5 was the first; 7 was the second; 11 was the third; 13 was the fourth; etc. These terms appeared in his popular 1557 mathematics text "The Whetstone of Witte" where, referring to the eighth power, Recorde wrote it "*doeth represent the square of squares squaredly*" (see Figure 1.17).

In Book 1 of his cardinal text *La* Géométrie, the famous French Mathematician Rene Descartes (1596–1650) introduced a close form of our present exponential notation by using superscripts. This notation replaced Recorde's traditional cossic notation used in early basic algebra. Some mathematicians, notably Isaac Newton, used exponents only for powers greater than 2. They represented squares as repeated multiplication and would write polynomials as:

$$ax + bxx + cx^3 + d$$

In 1748, the renowned Swiss mathematician Leonhard Euler (1707–1783) made a "quantum" exponential leap by considering

A Table of the Coffical Characters used in this Book.	Indices.	Characters.	Signification of the Characters.
	0	N	An Abfolute Number, as if it had no Mark.
	1	℞	The Root of any Number.
	2	3	A Square.
	3	℘	A Cube.
	4	33	A Squared Square, or Zenzizenzike.
	5	℞	A Surfolide.
	6	3℘	A Squared Cube, or Zenzicube.
	7	B℞	A Second Surfolide.
	8	333	A Zenzizenzizenzike, or Square of Squared Square.
	9	℘℘	A Cubed Cube.
	10	3℞	A Square of Surfolids.
	11	C℞	A Third Surfolide.
	12	33℘	A Zenzizenzicube, or Square of Squared Cubes.
	13	D℞	A Fourth Surfolide.
	14	3B℞	A Square of Second Surfolids.
	15	℘℞	A Cube of Surfolids.
	16	3333	A Zenzizenzizenzizenzike, or Square of Squares Squaredly (Squared.
	17	E℞	A Fifth Surfolide.
	18	3℘℘	A Zenzicubicube, or Square of Cubick Cubes.
	19	F℞	A Sixth Surfolide.
	20	33℞	A Square of Squared Surfolids.
	21	℘B℞	A Cube of Second Surfolids.
	22	3C℞	A Square of Third Surfolids.
	23	G℞	A Seventh Surfolide.
	24	333℘	A Square of Squares of Squared Cubes, or a
	&c.	&c.	Zenzizenzizenzicube.

Figure 1.17   Page from *The Whetstone of Witte*, 1557

functions in which the exponent was the variable, such as: $b^x = 53$. These were not algebraic functions, but were the lofty transcendental functions, which laid the groundwork for the exponential function e^x. The famous physicist Richard Feynman (1918–1988) called e^x a jewel and $e^{\pi i} = -1$ "the most remarkable formula in mathematics" (see page 370).

UNIT 9

The Origin of the Square-Root Symbol

When we use the square root symbol today, it is as common as any other mathematical symbol and yet, we don't question its origin. Read on and find out how this came to be. Although the concept of the square root is also shown on the Rhind Papyrus (1650 BCE) it was also already known to the ancient Indians and Chinese. Along with the first appearance in 1202 of our common numerals on the European continent in Fibonacci's book *Liber Abaci,* we also notice that he referred to the square root of a number with a capital letter R referring to the word radix in his book *Practica Geometriae* (1220). At one point of time, there was a small slash through the bottom right leg of the letter R, which could be confused with Px. As was mentioned earlier, Fibonacci adopted these symbols from his experiences in the Arabian world, largely from the mathematicians al-Khwarizmi and Omar Khayyam, who most likely got them from the Indian culture.

The square root symbol that we use today first appeared in 1525. We see this in a book, with the abbreviated title *Coss,* by Austrian mathematician Christoff Rudolff (1499–1545), which is believed to have been the first algebra textbook in the German language. In Figure 1.18, we show a page of this book where Rudolff used the square-root symbol for the first time.

After the second edition of this book in 1553, the square root symbol that he introduced became far more popular. Rudolff mentions "that the radix quadrata is, for brevity, designated in this algorithm with the symbol $\sqrt{}$, as$\sqrt{}4$" without the horizontal bar. The Swiss mathematician Leonhard Euler (1707–1783) felt that the symbol came from the handwritten version of the lowercase letter r. The horizontal-bar portion of the radical sign, which is used today was first introduced by the French mathematician René Descartes in his book *Geometrie* (1637). To indicate the cube root of a number Descartes used the following symbol $\sqrt{C}$, such as $\sqrt{C}27 = 3$, to

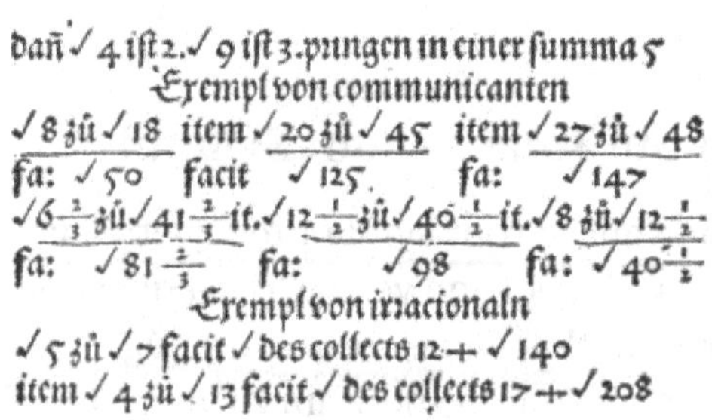

dañ $\sqrt{}$ 4 ift 2. $\sqrt{}$ 9 ift 3.prungen in einer fumma 5
Exempl von communicanten
$\sqrt{}$ 8 zü $\sqrt{}$ 18 item $\sqrt{}$ 20 zü $\sqrt{}$ 45 item $\sqrt{}$ 27 zü $\sqrt{}$ 48
fa: $\sqrt{}$ 50 facit $\sqrt{}$ 125. fa: $\sqrt{}$ 147
$\sqrt{}$ 6 ⅔ zü $\sqrt{}$ 41 ⅓ it. $\sqrt{}$ 12 ⅓ zü $\sqrt{}$ 46 ⅓ it. $\sqrt{}$ 8 zü $\sqrt{}$ 12 ½.
fa: $\sqrt{}$ 81 ⅓ fa: $\sqrt{}$ 98 fa: $\sqrt{}$ 40 ½
Exempl von irracionaln
$\sqrt{}$ 5 zü $\sqrt{}$ 7 facit $\sqrt{}$ des collects 12 + $\sqrt{}$ 140
item $\sqrt{}$ 4 zü $\sqrt{}$ 13 facit $\sqrt{}$ des collects 17 + $\sqrt{}$ 208

Figure 1.18

distinguish it from the numbers that belong to the square root symbol. Rudolff represented the cube root as $c\sqrt{}$ and the fourth root as $\sqrt{\sqrt{}}$, although each time without the horizontal bar. These symbols were used for many years by European mathematicians before they were finally adopted as we use them today. As you can see, there is a curious history as to how we came upon the square root symbol that we use today, when we write $\sqrt{25} = 5$.

UNIT 10

The Origin of Percent %, Number #, and Dollar $ Symbols

Have you ever pondered where the following signs came from: The Percent Symbol, %, the Number Symbol, #, and the Dollar sign $? These symbols evolved through several forms and historical contexts before they emerged as we know them today.

Percent Symbol %

Since percent is a ratio, or fraction, whose denominator is 100, you can see the reason for the fraction slash and the two zeroes, %, but that was not always the case. The percent sign is based on a power of 10 from our decimal system, however early number systems were not necessarily decimal systems.

The earliest known use of fractions whose denominator was 100 was in ancient Rome, long before the existence of the decimal system. The first Emperor of the Roman Republic, Augustus (63 BCE–14 CE), levied a tax of $\frac{1}{100}$ on goods sold at auction, known as *centesima rerun venalium*. Computation with these fractions was equivalent to computing percentages. As denominations of money grew in the Middle Ages, multiples of $\frac{1}{100}$ became more standard and more common in arithmetic books toward the end of the 15th century. By the 17th century, it was considered standard to quote interest rates in hundredths.

The first known use of a *percent* concept was in 1425, and it derived from the Latin *per centum*. This then transformed into the Italian term *per cento*, meaning "for a hundred". The "per" was often shortened to "p." and gradually disappeared completely. The "cento" morphed into two circles separated by a horizontal line, and which eventually evolved into the modern symbol "%".

Number Symbol

The number symbol, #, has a more colorful history and is known variously as Octothorpe, Pound, Grid, Diamond, Crosshatch, Hash, Tic-Tac-Toe, etc. This host of names belies its many uses throughout the centuries. We know it derives from the Roman term *libra pondo*, which means "pound in weight," which contracted over time to "lb" for *libra*. In the late 1600's, the famous English physicist and mathematician, Isaac Newton (1642–1727) employed the lb. contraction (see Figure 1.19), which inspired English printers to make it an official character and preserve it as a printing symbol. Eventually, English speakers put a line across the top to indicate the abbreviation and then hurried handwriting over time created the symbol "#." Examples of it being used to indicate pounds can be found at least as far back as 1850.

The earliest use of the symbol as a number sign can be found in the United States, where it is described in the 1853 bookkeeping treatise "Practical book-keeping" by John Coley Smith as the "number" character.

Its double meaning as a "number sign" (written before a figure) and a "pound sign" (written after a figure) is described in the instruction manual of the Blickendorfer model-5 typewriter from 1896. It later found its way onto 19th century typewriters as a "number mark" key. The QWERTY design of the keyboard of a typewriter was an American invention which was developed during the 1870s to avoid having letters frequently used consecutively not place side-by-side on the keyboard to avoid typing errors. In 1968, it was the touch tone telephone, created by Bell Labs in New Jersey that introduced the

Figure 1.19   The abbreviation written by Isaac Newton, showing the evolution from "℔" toward "#"

number key (#) to the fingers of the American public (often referred to as the "pound sign"). It was only occasionally used by frustrated callers who were manipulated by recorded voices, and today by computerized voices. Curiously, it also appeared on most Apple computer keyboards to only be coaxed up by hitting *Alt-3*. It certainly is a very versatile creation, not only indicating numbers, but also indicating weight, a musical sharp, three symbols ("###") indicating the end of a press release, also indicating the end of a mathematical proof, and as a space used by editors, to name just a few applications. Today, it has taken on great global prominence as the ubiquitous hashtag symbol on Twitter and can be found throughout world-wide media.

Dollar Sign $

The origin of the well-known United States dollar sign we would like to believe evolved from placing the U over the S, to get the symbol $, which, over time simplified to a single vertical bar, giving us the simple $. As nice as this story is, it is felt that this is more of a myth than truth. What is more likely is that in the 16th century, there was a great deal of commerce between Spanish explorers who discovered silver mines in the Americas and created coins called "pieces of eight", or "peso de ocho." Eventually, this abbreviated to "pesos," which during that time became the primary point for international trade. It was symbolized as a P with the superscript of S as P^s, and which further morphed into a P superimposed over the S, where only the vertical part of the P remained over the S giving us the common symbol for the dollar which we know as $. Since this predates the term "United States" we tend to believe that the original super imposition of the U on the S is merely a myth.

While we are on the topic of the dollar sign, $, it would be useful to know where the name "dollar" came from. It evolved from a large silver coin, originated in 1518, which stemmed from a Bohemian mine in Joachim's valley, which is in today's Czech Republic near the famous town of Karlovy Vary. In the 16th century German was the dominant language and the town was called Sankt Joachimsthal, which translates Saint Joachim's Valley. Taking the last part of Joachimsthal, that is, thal, the coin was referred to as a Thaler. This eventually resulted in the word dollar.

UNIT 11

The Ubiquitous Number π

Perhaps one of the most ubiquitous symbols in mathematics is the π symbol. When people recall their school mathematics instruction, the first two formulas that come to mind are those for the area of a circle (πr^2) and for the circumference of a circle ($2\pi r$), where r represents the radius of a circle.

While for some people π is nothing more than a touch of the button on their pocket calculator, where then a particular number appears on the readout, for others this number holds an unimaginable fascination. Depending on the size of the calculator's display, the number shown will be

    3.1415927, or
    3.141592654, or
    3.14159265359, or
    3.14159265358979323846264338327950, or even longer.

This push of a button still doesn't tell us what π actually is. Specifically, π was chosen to represent *the ratio of the circumference of a circle to its diameter.* This would be expressed symbolically as $\pi = \frac{C}{d}$, where C represents the length of the circumference and d represents the length of the diameter. The diameter of a circle is twice the length of the radius, $d = 2r$, where r is the length of the radius. If we substitute $2r$ for d we get $\pi = \frac{C}{2r}$, which leads us to the famous formula for the circumference of a circle: $C = 2\pi r$. As mentioned earlier, the other familiar formula containing π is that the area of a circle is πr^2.

You may be wondering by now where mathematicians actually got the idea to represent the ratio of the circumference of a circle to its diameter with the Greek letter π. According to the well-known Swiss-American mathematics historian, Florian Cajori (1859–1930), the symbol π was first used in mathematics by the English mathematician William Oughtred (1574–1660) in 1652 when he referred to the ratio of the circumference of a circle to its diameter as $\frac{\pi}{\delta}$, where π represented the periphery (perifereia), meaning perimeter, or the circumference of a circle[6] and δ represented the diameter.

In 1665, John Wallis used the Hebrew letter מ (mem), to equal one-quarter of the ratio of the circumference of a circle to its diameter (what, today, we would refer to as $\frac{\pi}{4}$). In 1706, William Jones (1675–1749) published his book, *Synopsis Palmariorum Matheseos*, where he used π to represent the ratio of the circumference of a circle to its diameter. This is believed to have been the first time that π was used as it is defined today. Jones' book alone would not have made the use of the Greek letter π to represent this geometric ratio as popular as it has become today. It was the legendary Swiss mathematician, Leonhard Euler (1707–1783), often considered the most prolific writer in the history of mathematics, who is largely responsible for today's common use of π. In 1736, Euler began using π to represent the ratio of the circumference of a circle to its diameter. But not until he used the symbol π in 1748 in his famous book *Introduction in Analysin Infinitorum* did

---

[6] Note, this is *not* what π later represented.

the use of π to represent the ratio of the circumference of a circle to its diameter become widespread.

The Area of a Circle Formula

Let's consider a relatively simple "derivation" of the formula $A = \pi r^2$ for the area of a circle with radius r. We begin by drawing a convenient size circle on a piece of cardboard. Divide the circle (which consists of 360°) into 16 equal arcs, as we show in Figure 1.20. This may be done by marking off consecutive arcs of 22.5° or by consecutively dividing the circle into two parts, then four parts, then bisecting each of these quarter arcs, and so on.

The 16 sectors we have constructed (shown in Figure 1.20) are then to be cut apart and placed in the manner shown in Figure 1.21.

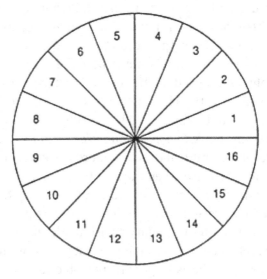

Figure 1.20

This placement suggests that we have a figure that approximates a parallelogram. That is, were the circle cut into more sectors, then the figure would look even more like a true parallelogram. Let us assume it is a parallelogram. In this case, the base would have a length of half

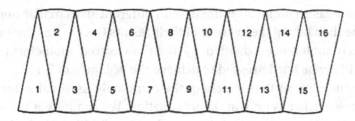

Figure 1.21

the circumference of the original circle, since half of the circle's arcs are used for each of the two sides of the approximate parallelogram. In other words, we formed something that resembles a parallelogram where one pair of opposite sides are not straight lines, rather they are circle arcs. We will progress as though they were straight lines, realizing that we will have lost some accuracy in the process, however, the more sectors that a circle is divided into, the closer the "parallelogram" will be to having a straight-line side. Therefore, the length of the base is $\frac{1}{2}C$. Since $C = 2\pi r$, the base length is, therefore, πr. The area of a parallelogram is equal to the product of its base and altitude. The altitude of this "parallelogram" is actually the radius r of the original circle. Therefore, the area of the "parallelogram" (which is actually the area of the circle we just cut apart) is $(\pi r) \times (r) = \pi r^2$, which gives us the commonly known formula for the area of a circle. For some readers, this might be the first time that the famous formula for the area of a circle, $A = \pi r^2$, has some real meaning.

The Value of π

Now that we have an understanding of what π meant in the context of these familiar formulas, we shall explore what is the actual value of π. One way to determine this ratio would be to carefully measure the circumference of a circle and its diameter and then find the quotient of these two values. This might be done with a tape measure or with a piece of string. An extraordinarily careful measurement might yield 3.14, but such accuracy is rare. As a matter of fact, to exhibit the difficulty of getting this two-place accuracy, imagine 25 people carrying out this measurement experiment with different size circular objects. Imagine then taking the average of their results (i.e. each of their measured circumferences divided by their measured diameters). You would likely be hard pressed to achieve the accuracy of 3.14.

You may recall that in school the commonly used value for π is 3.14 or $\frac{22}{7}$. Either is only an approximation. We cannot get the exact value of π. So how does one get a value for π? This has challenged mathematicians for centuries. With the aid of a computer Emma Haruka Iwao (1984–), a Japanese software engineer and computer scientist calculated the value of pi to 31.4 trillion digits (31,415,926,535,897) on March 14, 2019, which in the United States is considered Pi-day, since is written as 3-14.

Readers who wish to pursue the topic further are referred to *π: A Biography of the World's Most Mysterious Number,* A.S. Posamentier & I. Lehmann, Prometheus Books, 2004.

UNIT 12

One of the Earliest Approximations of π

You may be surprised to know that in biblical times, scholars believed that 3 was the value of π. For many years virtually all the books on the history of mathematics stated that in its earliest manifestation in history, namely in the Old Testament of the Bible, the value of π is given as 3. Yet recent "detective work" shows otherwise.[7]

One always relishes the notion that a hidden code can reveal long lost secrets. Such is the case with the common interpretation of the value of π in the Bible. There are two places in the Bible where the same sentence appears, identical in every way except for one word, which is spelled differently in the two citations. The description of a pool, or fountain, in King Solomon's temple is referred to in the passages that may be found in 1 *Kings* 7:23 and 2 *Chronicles* 4:2, and reads as follows:

---

[7]Alfred S. Posamentier and Noam Gordon, "An Astounding Revelation on the History of π" *Mathematics Teacher*, Vo. 77, No. 1, Jan. 1984, p.52.

> **"And he made the molten sea**[8] **of ten cubits from brim to brim,
> round in compass, and the height thereof was five cubits; and *a
> line* of thirty cubits did compass it round about."**

The circular structure described here is said to have a circumference of 30 cubits[9] and a diameter of 10 cubits. From this we notice that the Bible has $\pi = \frac{30}{10} = 3$. This is obviously a very primitive approximation of π. A late 18th century Rabbi, Elijah of Vilna[10] (1720–1797), was one of the great modern biblical scholars, who earned the title "Gaon of Vilna" (meaning brilliance of Vilna). He came up with a remarkable discovery, one that could make most history of mathematics books faulty, if they say that the Bible approximated the value of π as 3. Elijah of Vilna noticed that the Hebrew word for "line measure" was written differently in each of the two Biblical passages mentioned above.

In 1 *Kings* 7:23 it was written as קוה, whereas in 2 *Chronicles* 4:2 it was written as קו. Elijah applied the ancient biblical analysis technique (still used by Talmudic scholars today) called *gematria*, where the Hebrew letters are given their appropriate numerical values according to their sequence in the Hebrew alphabet, to the two spellings of the word for "line measure" and found the following. The letter values are: ק = 100, ו = 6, and ה = 5. Therefore, the spelling for "line measure" in 1 *Kings* 7:23 is קוה = 5 + 6 + 100 = 111, while in 2 *Chronicles* 4:2 the spelling קו = 6 + 100 = 106. Using gematria in an accepted way, he then took the quotient of these two values: $\frac{111}{106} = 1.0472$ (to four decimal places), which he considered the necessary "correction factor." By multiplying the Bible's apparent value (3) of π by this "correction factor," one gets 3.1416, which is π correct to four decimal places! "Wow!" Is a common reaction. Such accuracy is

---

[8] The "molten sea" was a gigantic bronze vessel for ritual ablutions in the court of the First Temple (966–955 B.C.). It was supported on the backs of 12 bronze oxen (volume $\approx$ 45,000 liters).

[9] A cubit is the length of a person's fingertip to his elbow.

[10] In those days Vilna was in Poland, while today the town is named Vilnius and is in Lithuania.

quite astonishing for ancient times. Moreover, just getting $\pi = 3.14$ using string measurements around a circle is quite a feat. Now imagine getting π accurate to four decimal places. We would contend that this would be nearly impossible with typical string measurements because of the accuracy needed in the measurement.

UNIT 13

The Origin of the Basic Trigonometric Names: Sine, Tangent, and Secant

Almost every high school student who took trigonometry remembers the mnemonic: S-OH C-AH T-OA, which stands for "sine — opposite over hypotenuse, cosine — adjacent over hypotenuse and tangent — opposite over adjacent." Were you one of the majority of students, who did not question where the names for these mathematical functions come from, or did you ponder their origin? The designations *tangent* and *secant* have clear mathematical origins, but *Sine* does not. Consider the term tangent first.

Figure 1.22 shows a unit circle where the radius $AC = 1$. The tangent of angle CAB is defined as the opposite side divided by the adjacent side or $\frac{BC}{AC} = \frac{BC}{1} = BC$. But observe that the line BC is a line tangent to the circle, hence the name *tangent* is used for this trigonometric function. While considering this diagram, we can also see where the term *secant* comes from. The secant of angle CAB is defined as the

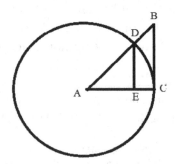

Figure 1.22   Unit Circle $AC = 1$

reciprocal of the cosine, and therefore, is the hypotenuse divided by the adjacent side or $\frac{AB}{AC} = \frac{AB}{1} = AB$. But the line AB is a secant line to the circle, which is a line that intersects a circle, and therefore the name *secant* is used for this trigonometric function. The word tangent comes from the Latin verb *tangere* which means "to touch" and the word secant comes from the Latin word *secare*, meaning "to cut". Both terms were first used by the Danish mathematician Thomas Fincke (1561–1656) in *"Geomietria Rotundi"* (1583).

The word *sine* has a more involute origin. It comes from the Latin for *sinus* which means bosom or fold. However, it evolved through a series of mistranslations starting with the Sanskrit word for half the chord: *jya-ardha*. Note, in Figure 1.22 the sine of angle *CAB* is defined as the opposite side divided by the hypotenuse or $\frac{DE}{AD}$. But AD is the radius = 1, so $\frac{DE}{AD} = \frac{DE}{1} = DE$, which is half the chord that extended would touch the circle at the other end. Hence the original Sanskrit word correctly identifies the meaning of sine. What happened next is due to the Indian mathematician Aryabhata (476–550) who abbreviated the Sanskrit word to *jya*. When *jya* was translated into Arabic, it incorrectly became *jiba*, which is not an Arabic word. Since Arabic is written without vowels, it was transcribed to *jb* causing later writers to interpret the consonants as *jaib*, which means bosom or breast. When an Arabic work in trigonometry was translated into Latin during the 12th century, the word *jaib* became the equivalent Latin word *sinus* which developed into our English word "sine."

Associated with the three basic trigonometric functions: sine, tangent and secant, are the cofunctions: cosine, cotangent and cosecant. The cofunctions are, respectively, reciprocals of the three basic functions. When one first encounters them, one may wonder why we need them. Their use in basic trigonometry is not very evident, however in more advanced mathematics, such as calculus, they aid in simplifying complex expressions and solving certain problems.

To complete the etymology involving the right triangle, the word hypotenuse, which is the side of the right triangle opposite the right angle, also comes from the Latin *hypotenusa*, which means "stretching under" (the right angle). This envisions the hypotenuse as the base of the triangle and the right angle above it. Most mathematical terms do tend to have logical origins. However, as we see here, occasionally, some mathematical meanings get lost in translation, or could possibly be a result of someone's imagination, and we find that they have little or no connection to what they represent.

UNIT 14

How Numbers are Categorized

Based on their properties, numbers fall into various categories. There are natural numbers, integers, rational numbers, irrational numbers, real numbers, imaginary numbers and complex numbers. The *natural numbers* are simply the positive whole numbers 1, 2, 3... etc. They are called the natural numbers because they are directly inspired by the natural world around us. They are the first number concepts found in a primitive civilization, or the first number concepts which a very young child grasps the meaning of. Years ago, anthropologists study-ing primitive civilizations (there are few, if any left now) found that some civilizations had only an understanding of the natural numbers 1, 2 and 3. After that it was the notion of "many."

Moving up the ladder of numbers, the *integers* include the natural numbers, the negative whole numbers, and zero. The *rational num-bers* are all the fractions, or ratios, and since the integers are fractions with denominators of 1, they are also rational numbers. The *irrational numbers* include numbers that are not ratios, such as square roots of numbers that are not perfect squares. For example, $\sqrt{3}, \sqrt{7}, \sqrt{21}$, etc., are irrational numbers. Also included are all roots of numbers that are not perfect roots, such as $\sqrt[3]{5}, \sqrt[5]{10}$, etc., as well as certain other num-bers such as π, which is the ratio of the circumference of a circle to its diameter, and e, which is a special limit that occurs in calculus. The *real numbers* consist of the rational *and* the irrational numbers, whereas the *imaginary numbers* are square roots of negative numbers that are not real such as $\sqrt{-1}, \sqrt{-4}$, etc. They were called imaginary because originally, they could not be put to any practical use which is not the case today. See the unit "Imaginary Numbers are not Really Imaginary" (page 40). The unit of the imaginary number system is

$\sqrt{-1}$, which is represented by the letter *i or j*. Finally, the *complex numbers* are numbers which have two parts: a real number combined with an imaginary number. Complex numbers include all the numbers mentioned above and no new numbers can be developed from the complex numbers. Any arithmetic or algebraic operation on a complex number, results in another complex number. Figure 1.23 shows each of these number groups and their hierarchy.

Considering Figure 1.23, if you move from the inside starting at natural numbers through the categories toward the outside, ending at complex numbers you would be almost tracing the historical development of numbers and much of mathematics. The earliest civilizations that we have any record of are the Sumerian, Babylonian and Egyptian several thousand years before the Common Era. These ancient civilizations were very mathematically advanced, but their number system included just the natural numbers and positive fractions, which sufficed for all their practical applications.

The next advance in the development of numbers came much later, after the Greek and Roman civilizations, as the natural numbers

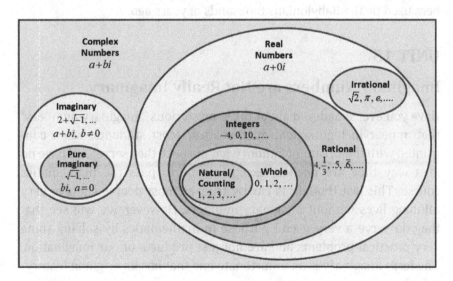

Figure 1.23   Number groups

and positive fractions sufficed for all the mathematics needed by these societies. The Greeks flirted with irrational numbers such as $\sqrt{2}$ and π, but could not come to terms with them, hence the word irrational came to mean unreasonable.

The concept of zero evolved in Indian mathematics around the 7th century (see the unit "The History of Zero" page 370). It was during the 9th century, with the significant contributions of Arabic mathematicians, including the introduction of algebra, that expanded the concepts of numbers. Algebraic theory grappled with positive and negative numbers, rational and irrational numbers, and treated them as "algebraic objects." These early beginnings are found in the work of the Islamic mathematician al-Khwarizmi (790–850). The word *algebra* comes from his name and the word *algorithm* from his work. The Arabs adopted the Indian numeral system, which is close to today's system, and it allowed for the expression of decimal fractions. The extraction of roots enabled a better understanding of irrational numbers and an expression for π. Number systems became more refined during the Renaissance when the concept of imaginary numbers was developed. Though it took many years and many civilizations to arrive at our present concepts of numbers, the reality is that most of us can function fine with only the numbers used by the Babylonians thousands of years ago.

UNIT 15

Imaginary Numbers are Not Really Imaginary

Have you ever wondered about those mysterious "imaginary numbers" that appear in high school mathematics? Most students, when confronted with them, cannot fathom what purpose they serve, and have no idea why they are needed, except maybe to pass the mathematics course. This last thought is clearly true, as most people can live very fulfilling lives without imaginary numbers. However, we will see that they do serve a very useful purpose in mathematics by solving some very practical problems and are not just products of our imagination. The term imaginary is an unfortunate one that has its origin in history.

Before we consider imaginary numbers, it is worth looking at what numbers really are. Observe that they are merely mental constructs that we have created to solve problems. They do not exist except in our minds. Some very early primitive civilizations did not have the concept of numbers, but visually they could distinguish between one thing, two things, three things and many things. Imaginary numbers are just that: a mental construct created to solve problems. What are these problems, you may ask? Well, imaginary numbers arose when mathematicians tried to make sense out of the square roots of negative numbers. As early as 50 CE, Heron of Alexandria (10 CE–70 CE) wrote that $\sqrt{81-144} = \sqrt{144-81} = \sqrt{63} = 8 - \frac{1}{16} = 7.9375$. Heron clearly did not accept the square root of a negative number. In fact, the answer in today's mathematics is approximately: $\sqrt{-63} = (\sqrt{63})(\sqrt{-1}) \approx 7.9373i$, where $i = \sqrt{-1}$ is the unit of the imaginary number system introduced by the Swiss mathematician Leonhard. Euler (1707–1783) in 1732.

For many years after Heron, square roots of negative numbers were not regarded as mathematical quantities. Though they do appear as roots of quadratic equations, they were first introduced in the solution of cubic equations by the Italian mathematician Gerolomo Cardano (1501–1576) during the Renaissance. The French mathematician René Descartes (1596–1650) is credited with applying the "unfortunate" term *imaginary* in 1637 to square roots of negative numbers in his work on geometry:

"For any equation one can imagine as many roots [as its degree would suggest], but in many cases no quantity exists which corresponds to what one imagines."

For example, for the simple quadratic equation: $x^2 + 9 = 0$, one "imagines" it has two roots. They are $\pm\sqrt{-9} = \pm3(\sqrt{-1}) = 3i$. Accepting imaginary numbers allows us to arrive at the *Fundamental Theorem of Algebra*, which is related to Descartes's quote above:

Every non-zero, single-variable, polynomial of degree n, with complex coefficients has, counted with multiplicity, exactly n complex roots.

In other words, a polynomial equation has as many roots as the highest power of the variable.

A complex number is one that has a *real* number (a rational or irrational number) combined with an imaginary number. Euler visualized complex numbers as points with rectangular coordinates, where one coordinate was the real number, and the other coordinate was the imaginary number. Euler is credited with the incredible formula which involves two transcendental numbers,[11] $e \approx 2.72$ and $\pi \approx 3.14$, and the imaginary unit $i = \sqrt{-1}$ with the equation $e^{i\pi} + 1 = 0$. See "Euler's Remarkable Relationship" (page 370).

Imaginary numbers did not receive universal acceptance until the 19th century. The great German mathematician Carl Friedrich Gauss (1777–1855) still believed, as late as 1825, that:

> *"The true metaphysics of the square root of –1 is elusive."*

Although complex numbers find uses in many branches of advanced mathematics, what we will show here is that during the late 19th century, a very practical use of complex numbers was discovered which simplified the solution of challenging problems in electricity. Until the late 19th century, DC (direct current) electricity was the main form of electricity in use and was generated from batteries. AC (alternating current) electricity slowly developed during the 19th century but was not a main form of power until 1884 when the Austrian-American engineer Nikola Tesla (1856–1943) invented the electric alternator, an electric generator that produces alternating current.

Let's first look at the most basic relationship in DC electricity. It is Ohm's law: $V = IR$
where V = voltage, I = current, and R = resistance. With this formula, one can solve a myriad of problems in DC electricity, because usually the current is constant, and its graph is simply a horizontal line. AC circuits are more complex than DC circuits. They are generated by a rotating coil of wire in a magnetic field and produce a wave of current

---

[11] A transcendental number is a number that is not algebraic — that is, not a root of a nonzero polynomial equation with integer or equivalently rational coefficients.

that looks just like a simple water wave. It is the wave produced by the sine function in mathematics (see Figure 1.24).

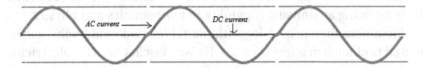

Figure 1.24   DC and AC currents

In the mid-19th century, the solution of AC circuit problems required vector analysis and calculus, which were advanced topics not generally taught to engineers at that time. Then along came the German-American mathematician and engineer Charles Proteus Steinmetz (Figure 1.25).

Figure 1.25   Albert Einstein and Charles Steinmetz

Charles Steinmetz was born in 1865 in Breslau, Germany with the physical deformity of a hunchback. At age 8, he struggled with multiplication tables. By age 10, he was one of the school's brightest pupils demonstrating an unusual capability in mathematics and physics.

Steinmetz memorized logarithmic tables and could solve exponential problems in a few seconds. He was fascinated with electricity, but his courses at the University of Breslau were lacking in electrical applications. He did not see a transformer until he came to the United States in 1889.

A custom official at Ellis Island refused him entry based on his lack of English, being a cripple, and having no money. A friend, Oscar Asmussen, agreed to take care of Steinmetz until he was on his own. The United States almost lost one of its most brilliant electrical engineers.

Steinmetz presented a simplified method of calculating values for alternating current using complex numbers and the "*j*-operator" at the International Electrical Congress in 1893. The *j*-operator was simply another notation for the imaginary number $i = \sqrt{-1}$. Complex numbers possessed a unique property of multiplication that mirrored the way alternating currents behave, and therefore, provided a much easier way to solve AC circuit problems. What follows is a simplified explanation of Steinmetz' method which you should be able to appreciate. Consider the repeated multiplication of the imaginary unit *i*:

$$i^0 = 1$$

$$i^1 = i$$

$$i^2 = -1$$

$$i^3 = (i^2)(i) = (-1)(i) = -i$$

$$i^4 = (i^2)(i^2) = (-1)(-1) = 1$$

$$...\text{(cycle repeats)}$$

After $i^4 = 1$, the powers of *i* repeat and the same four values, 1, *i*, −1, −*i*, keep occurring. This cyclic property is what makes imaginary numbers very different from real numbers. When you keep multiplying the unit of the real numbers, 1, by itself, the result is always 1. Now, let's look at the four imaginary products above as unit arrows or vectors on a graph, where the real numbers are plotted on the horizontal, or x-axis, and the imaginary numbers on the vertical, or y-axis (Figure 1.26). In Figure 1.26, $i^0 = 1$ is shown as an arrow that is drawn 1 unit long to the right, which is the positive horizontal direction. $i^1 = i$ is shown as an arrow that is drawn 1 unit long straight up, which is the positive vertical direction. Similarly, $i^3 = -i$ and $i^4 = -1$ are drawn as unit arrows in the opposite or negative directions.

The unique cyclic property that imaginary numbers possess causes the vectors on the graph to *rotate 90° each time you multiply by i*. This is what makes the algebra of imaginary numbers useful in analyzing and solving AC circuits. It is because certain electrical components in an AC circuit cause the current to speed up or slow down by a quarter of a cycle, which moves the path of the sine curve of the current to the right or to the left by 90°. This is equivalent to rotating the vectors on the graph by 90°. The two principal components that do this are capacitors and inductors. Capacitors, which consist of two parallel plates, store charge and cause the current to speed up by a quarter of a cycle. Inductors, which are coils, induce their own

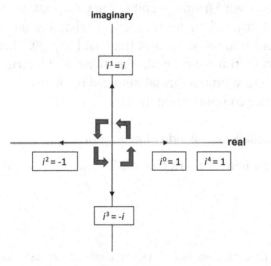

Figure 1.26

currents, and cause the current to slow down by a quarter of a cycle. Steinmetz was able to apply this cyclic property of imaginary numbers to AC circuits and greatly simplify the mathematical calculations. He, however, preferred to represent i as j, and called it the j-operator because it "operates" on the AC current and causes it to speed up or slow down by a quarter of a cycle.

Another interesting note about Ohm's Law $V = IR$, where V represents voltage, I represents current, and R represents resistance, and which simplifies calculations in DC circuits. There is an analogous law in AC circuits: $V = IZ$, where Z stands for reactance and is analogous to the. resistance R in a DC circuit and is a measure of the effect that AC components have on the current. Solving AC circuits using this formula was made much easier using complex numbers and the mathematics introduced by Charles Steinmetz.[12] We close here with a very entertaining anecdote about this brilliant man.

In 1903, General Electric encountered a problem with a huge electrical generator, which none of their engineers could solve. They brought in Steinmetz as a consultant. He found the problem difficult, but he closeted himself in with the generator for a few days and then chalk marked a large "X" on the casing. General Electric's engineers found a note telling them to cut the casing open at "X" and remove certain turns of wire from the stator. They did that and the generator worked perfectly. When he was asked what his fee would be, he thought about it for a while, and then said $1,000. This was a lot of money at that time! Stunned, the General Electric bureaucracy required him to submit a formal itemized invoice.

The invoice contained two items:

1. Marking chalk "X" on side of generator: $1

2. Knowing where to mark chalk "X": $999

---

[12]For more information on how complex numbers are used in electricity see "Mathematics for Electricity and Electronics", A. Kramer: Delmar, Fourth Edition 2012.

Charles Steinmetz's significant contribution took the "imaginary" out of imaginary numbers and made them an essential tool for solving electrical problems. So, if you studied or taught the topic of imaginary numbers, and questioned why they are needed, this should help to provide an answer and put that question to rest.

UNIT 16

The Birth of Algebra

Al–ge-bra, the word emerged in the title of the book written in 830 by the Persian mathematician and astronomer Muhammad ibn Musa al-Khwarizmi (ca. 780–850 CE). The treatise was called *"al-jabr wa'al-muqâbalah"* in Arabic and the word *al-jabr*, whose meaning is not totally clear, evolved into the English term algebra. Though al-Khwarizmi's work contains much of the basics of algebra studied in school today, its beginnings go much further back to ancient Egypt and the scribe Ahmes (ca. 1680–1620 BCE), the writer of the Rhind papyrus. The papyrus contained problems involving linear equations and series, but there was little evidence that algebra existed as a science. The ancient Greeks solved almost all their mathematical problems geometrically and they contain little resemblance to today's algebra. One major exception is that of Diophantus of Alexandria (ca. 208–292), sometimes known as the "father of Algebra." In his *Arithmetica* (ca. 275), we find the beginnings of algebraic symbolism, clear algebraic problems and solutions of equations solved by analytic methods (see Diophantine equations page 339). This work was the first devoted entirely to algebra, and his solution of indeterminate equations is today known as Diophantine analysis.

Exactly when algebra began in China is not easy to know but the work *Nine Sections* (ca. 1000 BCE) contains problems that can be solved by equations. The early Chinese did know how to solve quadratics during the 1st century BCE and in 250, Liu Hui (220–280) produced a series of rules which were tantamount to algebraic formulas though were in a rhetorical form. Chinese algebra culminated during the 13th century, when scholars solved higher-degree equations using methods which resemble ones developed in the Western world years later. In the near east country of India, we begin to see the roots of our western algebra. One of the earliest Hindu writers, the mathematician Brahmagupta (598–668) produced rules for solving quadratic equations and subsequent Hindu mathematicians dealt with problems involving series, radicals, and other types of equations.

We then come to the Arabs and the Persians who truly planted the roots of our modern system of algebra. Al-Khwarizmi, who was previously mentioned, is responsible for the name we use today and who was followed by the Persian mathematician Almâhâni (ca. 860) who worked on the cubic equation. Further contributions came from the Egyptian mathematician Abu Kamil (850–930) from whom Fibonacci drew material, the Persian mathematician al-Karkhi (953–1029) whose work *Fakhri* contains algebraic problems similar to problems appearing today and finally the Persian mathematician Omar Khayyam (1048–1131) who formulated some of the best algebra of the Persians.

The greatest western mathematician of the Middle Ages was Fibonacci (see page 333), whose *Liber Quadratorum* (ca. 1225) included solutions to this equation $x^2 + y^2 = z^2$ and other ingenious solutions of equations. Algebra flourished during the Renaissance and advanced far beyond the high school algebra of today.

UNIT 17

The Infinity Concept and Its Symbol ∞

One of the most curious and elusive concepts in mathematics is the measure of magnitude called *infinity*. A concept that only exists in our mind but escapes our reality. Thousands of years ago the ancient Greeks struggled with its meaning, and it remains as an idea not easily grasped by mathematicians and non-mathematicians alike. In the 5th century BCE, the Greek philosopher Zeno of Elea (ca. 495–ca. 430 BCE) was known for his paradoxes involving infinity which have confounded and challenged philosophers, mathematicians, and physicists for over two millennia. One of the more curious ones is that of Achilles, the great warrior, and the tortoise. Achilles starts running after the Tortoise, but every time he arrives at where the tortoise was, the Tortoise has moved on and is no longer there. The paradox is that

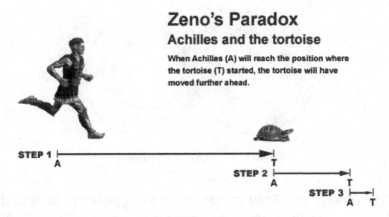

Figure 1.27   Achilles and the Tortoise

Achilles will never catch the Tortoise because it will take an infinite number of distances for Achilles to cover (see Figure 1.27).

Modern day calculus provides an answer to this sum of infinite distances by showing that the sum is not infinite but finite. For example, if the distances keep reducing by one-half and the first distance is one mile, the second distance $\frac{1}{2}$ mile, the third distance $\frac{1}{4}$ mile, etc., the infinite sum of the distances can then be shown mathematically to be:

$$1 + \frac{1}{2} + \frac{1}{4} + \frac{1}{8} + \cdots = 1 + 1 = 2$$

However, the paradox can still harbor uneasiness with one's reasoning. The infinity symbol ∞ has actually existed for millennia in various cultures with a myriad of meanings. In ancient Greece, it was known as the *Lemniscate* and derives from the Greek λημνίσκος and from the Latin "lēmniscātus" meaning "ribbons." It has its origins in the ancient symbol of a serpent biting its own tail which is the "ouroboros" from early Egyptian imagery and represents infinite immortality, self-fertilization and eternal emergence (see Figure 1.28). The infinity "ribbon" also finds its way into mystical Celtic knot designs which have no beginning or ending and hence is of infinite length.

Figure 1.28   Egyptian Ouroboros

The concept of infinity in mathematics can be traced back to the brilliant Greek mathematician and physicist Archimedes (288 BCE–212 BCE) around 250 BCE. Recently a modern X-ray scanner revealed his writings and work with finite sums that were concealed under some paintings and prayers. Prior to this discovery it was believed by most that the concept originated with Galileo (1564–1642). It was Galileo who helped to resolve some of the difficulties with infinite mathematical quantities that people struggled with. An outstanding paradox, sometimes called Galileo's paradox, has to do with the natural, or counting numbers, and their squares. On the surface it appears that, since most natural numbers are not perfect squares there should be more natural numbers than there are perfect squares of natural numbers.

However, since it is possible to show a one-to-one correspondence between each natural number and its square, that is, for every natural number there is a perfect square that it can be paired with, there must be as many infinite natural numbers as there are infinite perfect squares of natural numbers (see Figure 1.29).

$$
\begin{array}{ccccccc}
1 & 2 & 3 & 4 & 5 & 6 & 7 \ldots \\
\updownarrow & \updownarrow & \updownarrow & \updownarrow & \updownarrow & \updownarrow & \updownarrow \\
1 & 4 & 9 & 16 & 25 & 36 & 49 \ldots
\end{array}
$$

Figure 1.29   Natural numbers and their squares

Galileo realized that one cannot treat infinite quantities the same way one treats finite quantities. Problems arise only "when we attempt, with our finite minds, to discuss the infinite, assigning to it those properties which we give to the finite and limited; but this I think is wrong, for we cannot speak of infinite quantities as being the one greater or less than, or equal to another." "Furthermore, we can only infer that the totality of all numbers is infinite, and that the number of squares is infinite . . .; neither is the number of squares less than the totality of all numbers, nor the latter greater than the former; and finally, the attributes 'equal,' 'greater,' and 'less,' are not applicable to infinite, but only to finite quantities."

The infinity symbol appeared early in the Leviathan Cross, shown in Figure 1.30, which was adopted by the Knights Templar, a Catholic military order founded in 1119. It is also called the cross of Satan, because it was adopted by Anton LaVey (1930–1997), founder of the Church of Satan. On the bottom is the infinity sign which represents the eternal universe and infinity of nature. Above is a double cross which symbolizes balance and protection between people.

As a mathematical symbol, it first appeared in the 1655 work of the English mathematician John Wallis (1616–1703) "*De sectionibus conicis*", a treatise on conic sections. It is thought to have been inspired either by the Roman numeral for 1000, originally CIↃ, also CↃ, which was sometimes used to mean "many" or possibly the last Greek letter omega, ω. Wallis gave no clue as to why he adopted the symbol, however, it found its way into many applications symbolizing infinity and eternal limitlessness.

Figure 1.30   The Leviathan cross

Figure 1.31   Tarot card "The Juggler"

Figure 1.32   Symbol used by Euler to denote infinity

On the Tarot card known as The Juggler, shown in Figure 1.31, the infinity symbol appears as the Juggler's hat along with the mystical Jewish kabbalistic symbol aleph ℵ below the picture. Coincidentally, aleph was the symbol used by the German mathematician Georg Cantor (1845–1918) in the late 19th century as he developed the theory of the higher orders of infinity discussed below.

A variation of the symbol, shown in Figure 1.32, was employed by Swiss mathematician Leonhard Euler (1707–1783) to represent infinity. He performed a series of operations on infinity, which may have lacked rigor since calculus was not put on firm foundations until years later. The modern use of the symbol found everywhere today was well in place in the 18th century partly due to Euler's extensive use of the symbol.

Besides Zeno's paradoxes and Galileo's paradoxes, presented above, there are many other disquieting examples of infinite quantities that continue to play tricks on our psyche. A curious geometric one involves a staircase consisting of stairs of equal, or even unequal dimensions, as shown in the Figure 1.33.

The vertical rise of the staircase is denoted by a units, while the horizontal distance, traversed from the bottom of the first step Q to the top of the last step P, is denoted by b. Now suppose you want to carpet the stairs completely. Clearly you would need a length of carpeting equal to the sum of the bold segments comprising the steps as shown in the figure. This sum is equal to the total vertical distance plus the total horizontal distance, or $a + b$. Now suppose you decide to make the steps smaller which increases the number of stairs. You would still need the same amount of carpeting $a + b$ since the vertical

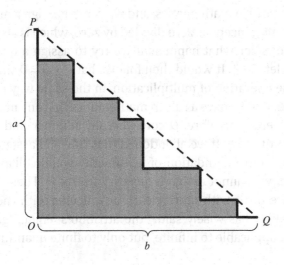

Figure 1.33   Carpeting stairs

and the horizontal distances remain the same. Now comes the enigma. Consider what happens if you keep increasing the number of steps making each one smaller and smaller. You would still need the same amount of carpeting, $a + b$, no matter how small you make the steps, since the vertical and horizontal distances do not change. But what happens when the number of steps increases without limit and the steps become infinitesimally small — so small that their dimensions approach zero in size. The total distance $a + b$ then appears to approach the length of the dashed line PQ shown in Figure 1.33. But PQ is the hypotenuse of the right triangle, and we know the hypotenuse of a right triangle is smaller than the sum of its two sides, since it is the shortest distance between the two points P and Q. Since PQ is equal to $\sqrt{a^2 + b^2}$ from the Pythagorean theorem, this relationship is expressed mathematically as: $\sqrt{a^2 + b^2} < a + b$, and the two quantities cannot be equal.

So, is there an explanation for this apparent paradox? It all stems from how we treat the concept of infinity. Two things are happening: As the size of the steps approach zero, the number of steps approach infinity. In the extreme case we have an infinite number of zero-line segments which make up the length PQ. We could then think of the length of PQ as being equal to zero times infinity. However, this is considered an indefinite form in mathematics, and we cannot assign a number to it. This is the same concept as zero divided by zero, which is also an indefinite form. Consider what happens if we try to assign a quantity to $\frac{0}{0}$. Suppose we let $Q = \frac{0}{0}$. It would then follow that $Q \times 0 = 0$ since division is the inverse operation of multiplication, in the same way that if $2 = \frac{6}{3}$, then $2 \times 3 = 6$. But we have a rule in mathematics that any number times zero equals zero. Therefore, Q could be any number and we cannot assign a quantity to $\frac{0}{0}$. If we abandoned that rule and assigned a quantity to $\frac{0}{0}$, our entire foundation of arithmetic would collapse! For the same reason, we cannot assign a quantity to $0 \times \infty$. These "apparent" paradoxes are dealt with more clearly in calculus with the theory of limits. As Galileo has wisely said: "the attributes 'equal,' 'greater,' and 'less,' are not applicable to infinite, but only to finite quantities."

What is perhaps, even more confounding, is the work of Georg Cantor (1845–1918). Cantor was born in St. Petersburg and developed a strong affection for culture and the arts. His mother, Maria Anna Böhm, was Russian, and being very talented musically, apparently fostered in him a gifted musical and artistic talent which led him to become an outstanding violinist. Later his family migrated to Germany and Cantor attended the University of Berlin where he collaborated with many of the leading German mathematicians. One of his outstanding contributions is related to Galileo's example, discussed above, that shows a one-to-one correspondence between the natural numbers and their squares. Cantor went much further and showed that there is a one-to-one correspondence between the natural numbers and the set of all fractions (rational numbers). This he called the first order of infinity, also called *countable infinity*, which he denoted by the ordinal number Aleph Null, $\aleph_0$, using the first letter of the Hebrew alphabet. In the early 1880's he went on to develop the foundations of set theory which became an important basis for what was called the "New Math" that was introduced in the United States school curriculum during the 1960's. Employing the concepts of set theory, Cantor proved that it is not possible to show a one to one correspondence between the set of all real numbers, consisting of the rational and irrational numbers, and the natural numbers. Hence the set of real numbers must constitute a higher order of infinity than the set of natural numbers whose cardinal number (i.e. the number of elements in the set), is sometimes referred to as the countable infinity. In other words, one can count the elements in the set in such a way that, even though the counting will take forever, eventually it is possible to count to any particular element in a finite amount of time. Cantor denoted by $\aleph_1$ the higher order of infinity of the real numbers whose nature is that it is uncountable. That is, the set is so large that we could never even count the elements between any two numbers in a finite amount of time no matter how close they were. These infinite concepts gave rise to the *continuum hypothesis*, which states that:

"There is no set whose cardinality is strictly between that of the integers and the real numbers."

The continuum hypothesis then poses the question as to whether there is a set whose elements are not countable and at the same time contain "fewer" elements than the real numbers. Suffice it to say that work on the continuum hypothesis has generated a lot of challenging mathematical and philosophical arguments and its proof remains a matter of interpretation today.[13]

The concept of infinity, no doubt, will remain an elusive and somewhat inexplicable one for some time, bridging the thoughts of mathematicians, philosophers, physicists and astronomers.

UNIT 18

The "Quad" in Quadratic

The "quad" in quadratic clearly stands for 4, however, quadratic in mathematics is used to identify an equation, function, or term of degree 2, that is x^2. When students are presented with the name of an object in mathematics, there is almost always an unconditional acceptance of the words used and rarely is the origin of the name questioned. Well, almost all mathematical names do have their origin in real experiences and calling a second-degree term x^2 quadratic is no exception. The general quadratic equation $ax^2 + bx + c = 0$, and the formula that is its solution:

$$x = \frac{-b \pm \sqrt{b^2 - 4ac}}{2a}$$

are among the few facts usually retained by students who are destined to take elementary algebra. However, have you ever asked yourself what the number 4 has to do with a term whose exponent is only a 2?

---

[13] See *Stanford Encyclopedia of Philosophy*, "The Continuum Hypothesis", May 2013

The mathematics of squaring a number goes back to antiquity, and the ancient Greeks grappled with its concepts, which included the square root of a number. Greek mathematics was heavily steeped in geometry and one of the impossible problems they devised was that of "squaring the circle", which meant geometrically constructing a square with the same area as that of a given circle by using only a finite number of steps with a pair of compasses and an unmarked straightedge.

The Greeks invented a term for the mathematicians who attempted to solve this problem, which meant "to busy oneself with the *quadrature*." The area of a square whose side is length x, is equal to x^2 and here we see the early use of the term quad referring to the area of a square because of its 4 sides. No doubt you can now see the justification for referring to x-to-the-second-power as "x squared" and using the word quadratic in this context. Quad comes from the Latin *quadratus*, which is the past participle of *quadrare* which means "to make square." Exactly when a second-degree term was called quadratic is not clear, but by the 17th century the name quadratic equation appears to describe an equation that involves the square and no higher powers of x. This should also enable us to justify why x^3 is called "x cubed," since a cube whose side-length is x will have its volume equal to x^3. This, however, is where the story ends as we cannot construct a figure in our world whose volume is equal to x^4.

UNIT 19
Prime Numbers: Background and Properties

Recall that a prime number is any natural number greater than one that has exactly two factors: the number 1 and the number itself. This makes the number 1 not a prime number since it does not have two factors. The concept of a prime number is essential to mathematics as it is an integral part of the *fundamental theorem of arithmetic*, which states that every natural number greater than 1 is either a prime number or can be expressed as the unique product of prime numbers. This theorem appears for the first time in approximately 300 BCE in Euclid's *Elements*. Although there are hints about this theorem around

1550 BCE in the Egyptian *Rhind Mathematical Papyrus*, we still credit Euclid as initiating the concept of prime numbers.

An early method for finding prime numbers was developed by the Greek mathematician Eratosthenes (276–194 BCE). Using his method, we list all the numbers beginning with the number 2 and cancel every successive multiple of 2. This leaves us with all the odd numbers. We then continue along with the next un-cancelled number, in this case, the number 3, and once again we cancel out all the multiples of 3. The next un-cancelled number is the number 5, and once again all multiples of 5 become cancelled. What remains un-cancelled, as we continue this process, are all the prime numbers. Figure 1.34 shows this "Sieve of Eratosthenes," a table of the odd numbers, with straight lines striking out the multiples of 3, 5, 7, 11, and 13. The remaining encircled numbers are the prime numbers between 3 and 281.

The Arabian mathematician Ibn al-Bannā' al-Marrākushī (1256–1321) discovered a method for moving more expeditiously along the Sieve of Eratosthenes by using only the divisors of the square root of the largest number to be considered.

Does the sequence of prime numbers continue indefinitely? Or will the process of cancelling multiples eventually stop, when all numbers have been struck out? In his book *Elements,* Euclid gave an ingenious proof that there are, indeed, infinitely many prime numbers. The argument is as follows: Suppose, we know only a finite number of prime numbers, say, for example, 2, 3, 5, 7, 11, 13, 17, and 19. Then we can show that there must be another one. That one would be found by forming the product of all known prime numbers, and adding 1 to obtain a larger number, which for our example would be as follows: $(2 \times 3 \times 5 \times 7 \times 11 \times 13 \times 17 \times 19) + 1 = 9,699,691$. This number turns out not to be a prime number because $347 \times 27,953 = 9,699,691$. If this number were a prime number, then we would have found another prime. The product of successive known primes plus 1 may be a prime, such as $(2 \times 3) + 1 = 7$, or may not be a prime, such as with $(3 \times 5) + 1 = 16$. Now assume that we did not generate a new

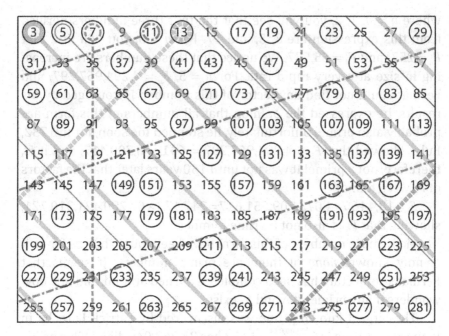

Figure 1.34   Sieve of Eratosthenes

prime — as was with the former case. Then we will look for the smallest number $q > 1$ that divides this new number exactly. The number q cannot be in our list of known prime numbers, because none of these divides 9,699,691 exactly (as there will always be a remainder 1). Then q must be a prime number itself, because otherwise q would be a product of smaller numbers, which are also factors of 9,699,691, and then q would not be the smallest factor. So, we have found a new prime number q that is not in our list of known prime numbers. For every finite list of prime numbers, we can find a new one, using this procedure. Therefore, the list of all prime numbers cannot be finite.

The famous French mathematician Pierre de Fermat (1601–1665) who made many significant contributions to the study of number theory, conjectured that all numbers of the form $F_n = 2^{2^n} + 1$, where $n = 0, 1, 2, 3, 4, \ldots$ were prime numbers. If you try this for F_n, where $n = 0, 1, 2$, you will see that the first three numbers derived from this expression are 3, 5, and 17. For $n = 3$, you will find that $F_n = 257$; and $F_4 = 65{,}537$. We notice that these numbers are increasing in size at a very rapid rate. For $n = 5$, $F_n = 4{,}294{,}967{,}297$, and Fermat could not find any factor of this number. Encouraged by his results, he expressed the opinion that all numbers of this form are probably also prime. Unfortunately, he stopped too soon, for in 1732, Euler showed that $F_5 = 4{,}294{,}967{,}297 = 641 \times 6{,}700{,}417$, and it is, therefore, not a prime! It was not until 150 years later that the factors of F_6 were found:

$$F_6 = 18{,}446{,}744{,}073{,}709{,}551{,}617 = 247{,}177 \times 67{,}280{,}421{,}310{,}721,$$

showing that F_6 is also not a prime number.

Many more numbers of this form have been found, but, as far as it is now known, *none* of them are prime numbers. It seems that Fermat's conjecture has been completely turned around, and one now wonders if any primes beyond F_4 exist.

The largest known prime was discovered by Patrick Laroche in January 2020 and is $2^{82{,}589{,}933} - 1$ and has 24,862,048 digits. This prime number is of the form $2^k - 1$, with some natural number k. Such primes are called *Mersenne primes* for the French monk Marin Mersenne (1588–1648), who developed the above formula and who created a list of Mersenne primes with exponents up to 257. The first few such Mersenne primes are obtained for $k = 2, 3, 5, 7, 13, 17, 19, 31, 61, 89, 107$, and 127.

You may have noticed that all the numbers k in this list are prime numbers themselves. Indeed, $2^k - 1$ can only be prime when k is prime. But this does not guarantee that when k is prime, $2^k - 1$ will also be prime. For example, $k = 11$ is prime, but $2^{11} - 1 = 2047 = 23 \times 89$ is not a prime, and so $k = 11$ does not qualify.

According to Euclid's proof, there are infinitely many prime numbers, although it is not known whether there are infinitely many Mersenne primes.

For some further insight into the prime numbers, we offer here a listing of all prime numbers less than 10,000 (see Figure 1.35).

2	3	5	7	11	13	17	19	23	29	31	37
41	43	47	53	59	61	67	71	73	79	83	89
97	101	103	107	109	113	127	131	137	139	149	151
157	163	167	173	179	181	191	193	197	199	211	223
227	229	233	239	241	251	257	263	269	271	277	281
283	293	307	311	313	317	331	337	347	349	353	359
367	373	379	383	389	397	401	409	419	421	431	433
439	443	449	457	461	463	467	479	487	491	499	503
509	521	523	541	547	557	563	569	571	577	587	593
599	601	607	613	617	619	631	641	643	647	653	659
661	673	677	683	691	701	709	719	727	733	739	743
751	757	761	769	773	787	797	809	811	821	823	827
829	839	853	857	859	863	877	881	883	887	907	911
919	929	937	941	947	953	967	971	977	983	991	997
1009	1013	1019	1021	1031	1033	1039	1049	1051	1061	1063	1069
1087	1091	1093	1097	1103	1109	1117	1123	1129	1151	1153	1163
1171	1181	1187	1193	1201	1213	1217	1223	1229	1231	1237	1249
1259	1277	1279	1283	1289	1291	1297	1301	1303	1307	1319	1321
1327	1361	1367	1373	1381	1399	1409	1423	1427	1429	1433	1439
1447	1451	1453	1459	1471	1481	1483	1487	1489	1493	1499	1511
1523	1531	1543	1549	1553	1559	1567	1571	1579	1583	1597	1601
1607	1609	1613	1619	1621	1627	1637	1657	1663	1667	1669	1693
1697	1699	1709	1721	1723	1733	1741	1747	1753	1759	1777	1783
1787	1789	1801	1811	1823	1831	1847	1861	1867	1871	1873	1877
1879	1889	1901	1907	1913	1931	1933	1949	1951	1973	1979	1987
1993	1997	1999	2003	2011	2017	2027	2029	2039	2053	2063	2069
2081	2083	2087	2089	2099	2111	2113	2129	2131	2137	2141	2143
2153	2161	2179	2203	2207	2213	2221	2237	2239	2243	2251	2267
2269	2273	2281	2287	2293	2297	2309	2311	2333	2339	2341	2347
2351	2357	2371	2377	2381	2383	2389	2393	2399	2411	2417	2423
2437	2441	2447	2459	2467	2473	2477	2503	2521	2531	2539	2543
2549	2551	2557	2579	2591	2593	2609	2617	2621	2633	2647	2657
2659	2663	2671	2677	2683	2687	2689	2693	2699	2707	2711	2713
2719	2729	2731	2741	2749	2753	2767	2777	2789	2791	2797	2801
2803	2819	2833	2837	2843	2851	2857	2861	2879	2887	2897	2903
2909	2917	2927	2939	2953	2957	2963	2969	2971	2999	3001	3011
3019	3023	3037	3041	3049	3061	3067	3079	3083	3089	3109	3119
3121	3137	3163	3167	3169	3181	3187	3191	3203	3209	3217	3221
3229	3251	3253	3257	3259	3271	3299	3301	3307	3313	3319	3323
3329	3331	3343	3347	3359	3361	3371	3373	3389	3391	3407	3413
3433	3449	3457	3461	3463	3467	3469	3491	3499	3511	3517	3527

Figure 1.35   Table of the first prime numbers under 10000

3529	3533	3539	3541	3547	3557	3559	3571	3581	3583	3593	3607
3613	3617	3623	3631	3637	3643	3659	3671	3673	3677	3691	3697
3701	3709	3719	3727	3733	3739	3761	3767	3769	3779	3793	3797
3803	3821	3823	3833	3847	3851	3853	3863	3877	3881	3889	3907
3911	3917	3919	3923	3929	3931	3943	3947	3967	3989	4001	4003
4007	4013	4019	4021	4027	4049	4051	4057	4073	4079	4091	4093
4099	4111	4127	4129	4133	4139	4153	4157	4159	4177	4201	4211
4217	4219	4229	4231	4241	4243	4253	4259	4261	4271	4273	4283
4289	4297	4327	4337	4339	4349	4357	4363	4373	4391	4397	4409
4421	4423	4441	4447	4451	4457	4463	4481	4483	4493	4507	4513
4517	4519	4523	4547	4549	4561	4567	4583	4591	4597	4603	4621
4637	4639	4643	4649	4651	4657	4663	4673	4679	4691	4703	4721
4723	4729	4733	4751	4759	4783	4787	4789	4793	4799	4801	4813
4817	4831	4861	4871	4877	4889	4903	4909	4919	4931	4933	4937
4943	4951	4957	4967	4969	4973	4987	4993	4999	5003	5009	5011
5021	5023	5039	5051	5059	5077	5081	5087	5099	5101	5107	5113
5119	5147	5153	5167	5171	5179	5189	5197	5209	5227	5231	5233
5237	5261	5273	5279	5281	5297	5303	5309	5323	5333	5347	5351
5381	5387	5393	5399	5407	5413	5417	5419	5431	5437	5441	5443
5449	5471	5477	5479	5483	5501	5503	5507	5519	5521	5527	5531
5557	5563	5569	5573	5581	5591	5623	5639	5641	5647	5651	5653
5657	5659	5669	5683	5689	5693	5701	5711	5717	5737	5741	5743
5749	5779	5783	5791	5801	5807	5813	5821	5827	5839	5843	5849
5851	5857	5861	5867	5869	5879	5881	5897	5903	5923	5927	5939
5953	5981	5987	6007	6011	6029	6037	6043	6047	6053	6067	6073
6079	6089	6091	6101	6113	6121	6131	6133	6143	6151	6163	6173
6197	6199	6203	6211	6217	6221	6229	6247	6257	6263	6269	6271
6277	6287	6299	6301	6311	6317	6323	6329	6337	6343	6353	6359
6361	6367	6373	6379	6389	6397	6421	6427	6449	6451	6469	6473
6481	6491	6521	6529	6547	6551	6553	6563	6569	6571	6577	6581
6599	6607	6619	6637	6653	6659	6661	6673	6679	6689	6691	6701
6703	6709	6719	6733	6737	6761	6763	6779	6781	6791	6793	6803
6823	6827	6829	6833	6841	6857	6863	6869	6871	6883	6899	6907
6911	6917	6947	6949	6959	6961	6967	6971	6977	6983	6991	6997
7001	7013	7019	7027	7039	7043	7057	7069	7079	7103	7109	7121
7127	7129	7151	7159	7177	7187	7193	7207	7211	7213	7219	7229
7237	7243	7247	7253	7283	7297	7307	7309	7321	7331	7333	7349
7351	7369	7393	7411	7417	7433	7451	7457	7459	7477	7481	7487
7489	7499	7507	7517	7523	7529	7537	7541	7547	7549	7559	7561
7573	7577	7583	7589	7591	7603	7607	7621	7639	7643	7649	7669
7673	7681	7687	7691	7699	7703	7717	7723	7727	7741	7753	7757

Figure 1.35   (*Continued*)

7759	7789	7793	7817	7823	7829	7841	7853	7867	7873	7877	7879
7883	7901	7907	7919	7927	7933	7937	7949	7951	7963	7993	8009
8011	8017	8039	8053	8059	8069	8081	8087	8089	8093	8101	8111
8117	8123	8147	8161	8167	8171	8179	8191	8209	8219	8221	8231
8233	8237	8243	8263	8269	8273	8287	8291	8293	8297	8311	8317
8329	8353	8363	8369	8377	8387	8389	8419	8423	8429	8431	8443
8447	8461	8467	8501	8513	8521	8527	8537	8539	8543	8563	8573
8581	8597	8599	8609	8623	8627	8629	8641	8647	8663	8669	8677
8681	8689	8693	8699	8707	8713	8719	8731	8737	8741	8747	8753
8761	8779	8783	8803	8807	8819	8821	8831	8837	8839	8849	8861
8863	8867	8887	8893	8923	8929	8933	8941	8951	8963	8969	8971
8999	9001	9007	9011	9013	9029	9041	9043	9049	9059	9067	9091
9103	9109	9127	9133	9137	9151	9157	9161	9173	9181	9187	9199
9203	9209	9221	9227	9239	9241	9257	9277	9281	9283	9293	9311
9319	9323	9337	9341	9343	9349	9371	9377	9391	9397	9403	9413
9419	9421	9431	9433	9437	9439	9461	9463	9467	9473	9479	9491
9497	9511	9521	9533	9539	9547	9551	9587	9601	9613	9619	9623
9629	9631	9643	9649	9661	9677	9679	9689	9697	9719	9721	9733
9739	9743	9749	9767	9769	9781	9787	9791	9803	9811	9817	9829
9833	9839	9851	9857	9859	9871	9883	9887	9901	9907	9923	9929
9931	9941	9949	9967	9973							

Figure 1.35　(*Continued*)

UNIT 20

Perfect Numbers

For quite a few years the European Union boasted that they had a perfect contingent of countries with 28 members. This was true until the United Kingdom withdrew from this perfect number of countries. The reason that we can refer to the original 28-member group as a perfect number of countries is that the number 28 is considered a perfect number in mathematics. While we might assume that everything in mathematics is perfect, might there still be anything more perfect than it? In the field of number theory, we have an entity called a "perfect number". This is defined as a number equal to the sum of its proper divisors (i.e. all the divisors except the number itself). The smallest perfect number is 6, since $6 = 1 + 2 + 3$, which is the sum of all its divisors excluding the number 6 itself.[14] The next larger perfect number is 28, since again $28 = 1 + 2 + 4 + 7 + 14$. It is then quite a long

Rank	k	Perfect number	Number of Digits	Year discovered
1	2	6	1	Known to the Greeks
2	3	28	2	Known to the Greeks
3	5	496	3	Known to the Greeks
4	7	8128	4	Known to the Greeks
5	13	33550336	8	1456
6	17	8589869056	10	1588
7	19	137438691328	12	1588
8	31	2305843008139952128	19	1772
9	61	265845599...953842176	37	1883
10	89	191561942...548169216	54	1911
11	107	131640364...783728128	65	1914
12	127	144740111...199152128	77	1876
13	521	235627234...555646976	314	1952
14	607	141053783...537328128	366	1952
15	1279	541625262...984291328	770	1952
16	2203	108925835...453782528	1327	1952
17	2281	994970543...139915776	1373	1952
18	3217	335708321...628525056	1937	1957
19	4253	182017490...133377536	2561	1961

Figure 1.36

---

[14] It is also the only number that is the sum and product of the same three numbers: $6 = 1 \times 2 \times 3 = 3!$. Also $6 = \sqrt{1^3 + 2^3 + 3^3}$. It is also fun to notice that $\frac{1}{1} = \frac{1}{2} + \frac{1}{3} + \frac{1}{6}$. By the way, while on the number 6, it is nice to realize that both 6 and its square, 36, are triangular numbers (see page 67).

20	4423	407672717...912534528	2663	1961
21	9689	114347317...429577216	5834	1963
22	9941	598885496...073496576	5985	1963
23	11213	395961321...691086336	6751	1963
24	19937	931144559...271942656	12003	1971
25	21701	100656497...141605376	13066	1978
26	23209	811537765...941666816	13973	1979
27	44497	365093519...031827456	26790	1979
28	86243	144145836...360406528	51924	1982
29	110503	136204582...603862528	66530	1988
30	132049	131451295...774550016	79502	1983
31	216091	278327459...840880128	130100	1985
32	756839	151616570...565731328	455663	1992
33	859433	838488226...416167936	517430	1994
34	1257787	849732889...118704128	757263	1996
35	1398269	331882354...723375616	841842	1996
36	2976221	194276425...174462976	1791864	1997
37	3021377	811686848...022457856	1819050	1998
38	6972593	955176030...123572736	4197919	1999
39	13466917	427764159...863021056	8107892	2001
40	20996011	793508909...206896128	12640858	2003
41	24036583	448233026...572950528	14471465	2004
42	25964951	746209841...791088128	15632458	2005
43	30402457	497437765...164704256	18304103	2005
44	32582657	775946855...577120256	19616714	2006
45	37156667	204534225...074480128	22370543	2008
46	42643801	144285057...377253376	25674127	2009
47	43112609	500767156...145378816	25956377	2008
48	57885161	169296395...270130176	34850340	2013

Figure 1.36 (*Continued*)

way to get to the next perfect number, which is $496 = 1 + 2 + 4 + 8 + 16 + 31 + 62 + 124 + 248$, as it is the sum of all of its divisors. The first four perfect numbers 6, 28, 496, and 8128 were known to the ancient Greeks. It was Euclid who came up with a theorem to generalize a procedure to find a perfect number, where for an integer k, if $2^k - 1$ is a prime number, then $2^{k-1} (2^k - 1)$ is a perfect number. We do not have to use all values of k, since if k is a composite number, then $2^k - 1$ is also a composite number[15]. Using Euclid's method for generating perfect numbers, we get the following table (Figure 1.36), where for the values of k, we get $2^{k-1} (2^k - 1)$, as perfect numbers when $2^k - 1$ is a prime number.

In Figure 1.36, the perfect numbers are abbreviated, however, below we show a few of these in their complete form:

2,305,843,008,139,952,128
2,658,455,991,569,831,744,654,692,615,953,842,176
191,561,942,608,236,107,294,793,378,084,303,638,130,997,321,54
8,169,216

By observation, we notice some properties of perfect numbers. They all seem to end in either a 6 or a 28, and these are preceded by an odd digit. They also appear to be triangular numbers (see page 67), which are the sums of consecutive natural numbers (e.g. $496 = 1 + 2 + 3 + \cdots + 29 + 30 + 31$).

---

[15] If $k = pq$, then $2^k - 1 = 2^{pq} - 1 = (2^p - 1)(2^{p(q-1)} + 2^{p(q-2)} + \ldots + 1)$. Therefore, $2^k - 1$ can only be prime when k is prime, but this does not guarantee that when k is prime, $2^k - 1$ will also be prime, as can be seen from the following values of k:

k	2	3	5	7	11	13
$2^k - 1$	3	7	31	127	2047	8191

where $2047 = 23 \times 89$ is not a prime and so doesn't qualify.

From the work of Italian mathematician Franciscus Maurolycus (1494–1575), we know that every *even* perfect number is a hexagonal number (1, 6, 15, 28, ...). In general, it is obtained by $k(2k-1)$.

To take it a step further, every perfect number after the number 6 is the partial sum of the series: $1^3 + 3^3 + 5^3 + 7^3 + 9^3 + 11^3 + \cdots$

For example,

$28 = 1^3 + 3^3$, and
$496 = 1^3 + 3^3 + 5^3 + 7^3$
$8{,}128 = 1^3 + 3^3 + 5^3 + 7^3\ 9^3 + 11^3 + 13^3 + 15^3.$

This connection between the perfect numbers greater than 6 and the sum of the cubes of consecutive odd numbers is far more than could ever be expected! You might try to find the partial sums for the next perfect numbers. We do not know if there are any odd perfect numbers, since none has yet been found.

UNIT 21

Triangular Numbers and Their Relation to Square Numbers

It is to be assumed that everybody knows what square numbers are, such as the numbers: 4, 9, 16, 25, 36, 49, 81, 100, 121, 144, ... They are typically called *square numbers,* since these numbers can be arranged in a square arrangement of dots as shown in Figure 1.37.

Another common shape for numbers is the triangle. In the days of Pythagoras, dots were arranged in a triangular shape for *triangular numbers* as shown in Figure 1.38:

While the Greek philosophers would not have included the number 1, it is included today in the list of *triangular numbers*, as it is included in the list of square numbers. So, the sequence of triangular numbers is 1, 3, 6, 10, 15, 21, 28, 36, 45, ...

There are clever ways to generate triangular numbers as we can see in Figure 1.39.

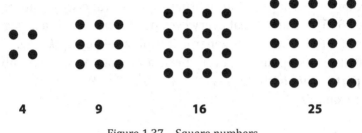

4   **9**   **16**   **25**

Figure 1.37   Square numbers

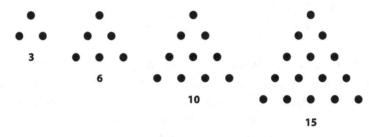

3   **6**   **10**   **15**

Figure 1.38   Triangular numbers

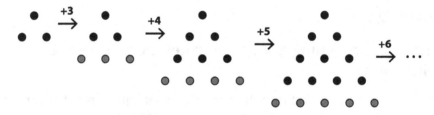

Figure 1.39   Generating triangular numbers

We obtain the next triangular number just by adding a row at the bottom of the previous triangle. Every row has one dot more than the previous row. So, the triangular numbers are simply created as sums of natural numbers, as we show below.

$$1 = 1,$$
$$1 + 2 = 3,$$
$$1 + 2 + 3 = 6,$$
$$1 + 2 + 3 + 4 = 10,$$
$$1 + 2 + 3 + 4 + 5 = 15,$$
and so on …

To generate the nth triangular number, which we conveniently call T_n, we simply find the sum of all natural numbers up to n (always including the number 1): $T_n = 1 + 2 + ... + n$, where n is any natural number.

For the Pythagoreans, this shape had a special meaning. It was called the "tetraktys" and seen as a divine symbol of perfection representing the whole cosmos, including the sum of all possible dimensions. The first row, a single point, is the unity that generates all other dimensions. With the two points in the second row, they believed that one can represent a one-dimensional line. The third row, which consists of three points, can be arranged as a triangle in a two-dimensional plane, and the fourth row, which has four points, can be arranged to outline a three-dimensional figure, namely a tetrahedron. The sum of all these is 10, the Dekad, which was also the base of the number system that was already in use in ancient Greece. In the Attic numeral system that was used by the Athenians in the 5th century BCE, the numeral for 10 was a Delta "Δ", which was the first letter of the word "Deka" (Δεκα) indicating "ten" and one can't help noticing the similarity between Δ and the triangular outline of the tetraktys.

There is a very nice relationship between triangular numbers and square numbers which we can best show by the transition in Figure 1.40. We simply add two consecutive triangular numbers and see that it forms a square number. For example, the sum of the two consecutive triangular numbers, 10 and 15 is 25, which is a square number as shown in Figure 1.40.

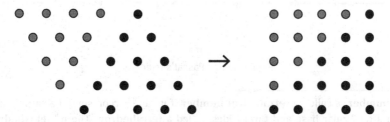

Figure 1.40   Sum of two Consecutive triangular numbers

We would write that algebraically as follows $T_{n-1} + T_n = n^2$ (for all natural numbers n greater than 1).

The triangular numbers tend to appear when they are least expected. For example, in the famous Pascal triangle, which we show in Figure 1.41, the first oblique line drawn from right to left contains all 1's and the next parallel oblique line is comprised of the natural numbers. We are presented with the 3rd such oblique line as the triangular numbers, which is shown in Figure 1.41.

We can even use Pascal's triangle to fetch various triangular numbers using an oblique line from left to right, as shown in Figure 1.42.

The ambitious reader may notice that the fourth parallel oblique line drawn from right to left contains the tetrahedral numbers,[16]

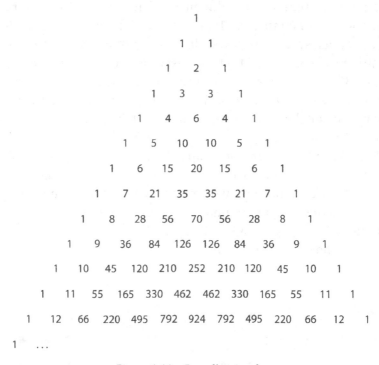

Figure 1.41   Pascal's triangle

---

[16] A **number** is called a **tetrahedral number** if it can be represented as a pyramid with a triangular base and three sides, called a **tetrahedron**. The nth **tetrahedral number** is the sum of the first n triangular **numbers**. The first nine **tetrahedral numbers** are: 1, 4, 10, 20, 35, 56, 84, 120, 165, ...

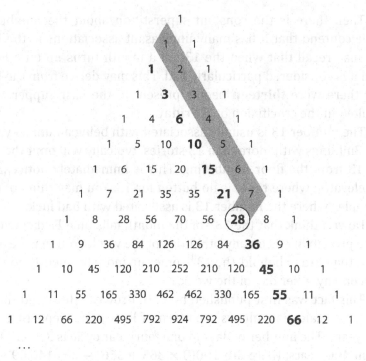

Figure 1.42   Triangular numbers in Pascal's triangle

which one could say are the three-dimensional analog of the triangular numbers.

UNIT 22

Numerology

Numerology studies numbers from non-mathematical points of view, such as psychology and philosophy. In 1995, a study[17] concluded that the easiest numbers to remember up to 100 were in priority order: 8, 1, 100, 2, 17, 5, 9, 10, 99, and 11 while the most difficult numbers to remember were: 82, 56, 61, 94, 85, 45, 83, 59, 41, and 79. This psychological study need not be relevant to you but may be for others!

---

[17] *Milikowski, Marisca; Elshout, Jan J.*: What makes numbers easy to remember? *British Journal of Psychology* 1995, p. 537–547.

Then there is the constant superstition about the number 13. Some contend that it has many unpleasant associations in the Bible. You may recall that when the 13th of a month turns up on a Friday, then it is considered particularly bad. This may derive from the belief that there were **thirteen** people present at the Last Supper, which resulted in the crucifixion on a **Friday.**

The number 13 is usually associated with being an unlucky number. Buildings with more than 13 stories, typically will omit the number 13 from the floor numbering. This is immediately noticeable in the elevator, where there is no button for 13. You may think of other examples where the number 13 is associated with bad luck.

Do you think that the 13th of the month falls on a Friday with the same probability as on any other day of the week? You may be astonished that, lo and behold, the 13th comes up more frequently on Friday than on any other day of the week.

This fact was first published by B. H. Brown.[18] He stated that the Gregorian calendar follows a pattern of leap years, repeating every 400 years. The number of days in one four-year cycle is $3 \times 365 + 366$. So, in 400 years, there are $100(3 \times 365 + 366) - 3 = 146,097$ days. Note that the century year, unless divisible by 400, is not a leap year, that is why we subtract 3. This total number of days is exactly divisible by 7. Since there are 4,800 months in this 400-year cycle, the 13th comes up 4800 times according to the table shown Figure 1.43. Interestingly enough, the 13th comes up on a Friday more often than on any other day of the week. Figure 1.43 summarizes the frequency

Day of the week	Number of 13s	Percent
Sunday	687	14.313
Monday	685	14.271
Tuesday	685	14.271
Wednesday	687	14.313
Thursday	684	14.250
Friday	*688*	*14.333*
Saturday	684	14.250

Figure 1.43

[18] "Solution to Problem E36." *American Mathematical Monthly*, 1933, vol. 40, p. 607.

of the 13th appearing on the days of the week. Albeit a slight difference and perhaps mathematically insignificant, it still allows Friday to stand out from the rest.

Then there are well-known triskaidekaphobics (which is a phobia for the number 13), such as Napoleon Bonaparte, Herbert Hoover, Mark Twain, Richard Wagner, and Franklin Delano Roosevelt. For example, Roosevelt never wanted to sit with thirteen people at dinner and he died on Thursday, April 12, 1945, possibly avoiding the next day: Friday the 13th! Consider Richard Wagner (1813–1883), the famous German composer who revolutionized music. For him, the number 13 has played a curious role:

Wagner was born in 18**13**, which has a digit sum of **13**.
Wagner died on February **13**, 1883, which was the **13**th year of the unification of Germany.
Wagner wrote **13** operas during his lifetime.
Richard Wagner's name consists of **13** letters.
Wagner's opera, Tannhäuser, was completed on April **13**, 1845.

On May **13**, 1861, Wagner premiered Tannhäuser in Paris during his **13**-year exile from Germany.

The grand opening of Wagner's festival opera house in Bayreuth, Germany was opened on August **13**, 1876.

Wagner completed his last opera, Parsifal, on January **13**, 1882.

Wagner's last day in Bayreuth, the city in which he built his famous festival opera house, which still flourishes today, was September **13**, 1882.

Who said numbers can't form beautiful relationships! Having experienced some of these unique situations might give you the feeling that there is more to "numbers" than meets the eye. While, rightfully, you may want to verify these number relationships to everyday life, we encourage you to find others that can also be considered "beautiful". This kind of "numerology" is just for entertainment and does not involve mathematics. There are certainly more fascinating numerical relationships, and we leave them for you to discover.

UNIT 23

Very Small and Very Large Numbers

One's daily needs do not usually encounter very small and very large numbers, however they do play an important role in our increasingly technical world. Recall the basic metric prefixes, shown in Figure 1.44, that are used to represent the powers of 10.

The positive and the negative powers of 10 in Figure 1.44 are a large part of electrical terminology such as megavolts, microamps, nanofarads (a small capacitor) etc., and the units kilogram, millimeter, and kilometer occur often. The positive powers of 10 shown in Figure 1.44 are most familiar as they are used often to describe the various storage capacities of computer hardware: megabytes (MB), gigabytes (GB), terabytes (TB) etc. Numbers larger than 10^{12} can occur in astronomy, mathematics, statistical mechanics and cryptography whereas numbers smaller than 10^{-12} can occur in mathematics, nano technology, atomic physics and microbiology.

However, it is interesting to explore the names that have been devised for very large and very small numbers as they have fascinated people as far back as Archimedes. Curiously, all countries have not equally applied the same logic for their names. Let us start with

Power of 10	Prefix	Abbreviation
10^{12}	tera	T
10^{9}	giga	G
10^{6}	mega	M
10^{3}	kilo	k
10^{-3}	milli	m
10^{-6}	micro	μ
10^{-9}	nano	n
10^{-12}	pico	p

Figure 1.44   Metric prefixes

the number one million, whose name is the only one that is universally accepted and has its origin in the early Italian word *millione* coined around the 15th century. The prefix *mille* comes from the Latin for one thousand and was combined with the suffix *one* to mean "great thousand." Figure 1.45 shows the names for large numbers up to decillion, and the differences between the names used by various countries.

The Latin prefixes bi, tri, quad etc. were first used by the French mathematician Nicholas Chuquet (1484–1488) for billion, trillion, quadrillion etc. In the Short Scale, the power of 10 is expressed in the form 10^{3n+3} where the Latin prefix for n is used to name the number. For example, for $10^{27} = 10^{3(8)+3}$, $n = 8$, and therefore, the prefix *oct* is assigned to the name. The Long Scale expresses the power of 10 in the form 10^{6n} and assigns the prefix based on the value of n. The exceptions are powers of 10 not divisible by 6 such as 10^9 and 10^{15}, which are named milliard and billiard, respectively. The term milliard is

Name	Short scale (US, English Canada, *modern* British, Australia, and Eastern Europe)	Long scale (French Canada, *older* British, Western & Central Europe)
Million	10^6	10^6
Milliard	*Not used*	10^9
Billion	10^9	10^{12}
Billiard	*Not used*	10^{15}
Trillion	10^{12}	10^{18}
Quadrillion	10^{15}	10^{24}
Quintillion	10^{18}	10^{30}
Sextillion	10^{21}	10^{36}
Septillion	10^{24}	10^{42}
Octillion	10^{27}	10^{48}
Nonillion	10^{30}	10^{54}
Decillion	10^{33}	10^{60}

Figure 1.45   Very large numbers

never used in the United States but is used often in continental Europe and in non-English speaking countries. Note that it is possible a "billionaire" in French Canada might be considered a trillionaire in English Canada!

Following this pattern, we have the largest named number for each scale, centillion, shown in Figure 1.46.

Name	Short scale	Long scale
Centillion	10^{303}	10^{600}

Figure 1.46   Largest named number

Two very large numbers, which are less than official, the googol = 10^{100} and the googolplex = 10^{googol}. They are discussed in the unit "The Googol is Much Older Than Google" on page 77.

Occasionally numbers larger than a trillion have occurred in everyday life as a result of runaway inflation. After World War II in 1946, Hungary printed the highest numbered bank note ever produced for one sextillion (10^{21}) pengõ. The large number names in Figure 1.45 are actually not employed in technical use, since Scientific Notation, which expresses a number using a power of 10, is more convenient to work with. For example, in the short scale, 5×10^{45} is more easily said then "Five Quattuordecillion". Most disciplines also prefer to use the metric prefixes in Figure 1.44 which go as far as 10^{24}(yotta). Astronomy has its own specific terms for large distances, such as a light year, the distance light travels in a year, which is equal to 9.461×10^{12} km, or almost 6 trillion miles, and a parsec which is equal to 3.26 light years.

The names for very small numbers are shown in Figure 1.47 and derive their designation from those for very large numbers with the diminutive suffix 'th' added.

Though we rarely encounter very large or very small numbers in our daily routine, they do possess an intellectual and mathematical fascination, and so we are motivated to provide them with an identity.

Power of ten	Short scale (U.S. and modern British)	Long scale (continental Europe, archaic British and India)	SI prefix	SI symbol
10^{-1}	Tenth	Tenth	deci-	d
10^{-2}	Hundredth	Hundredth	centi-	c
10^{-3}	Thousandth	Thousandth	milli-	m
10^{-6}	Millionth	Millionth	micro-	μ
10^{-9}	Billionth	Milliardth	nano-	n
10^{-12}	Trillionth	Billionth	pico-	p
10^{-15}	Quadrillionth	Billiardth	femto-	f
10^{-18}	Quintillionth	Trillionth	atto-	a
10^{-21}	Sextillionth	Trilliardth	zepto-	z
10^{-24}	Septillionth	Quadrillionth	yocto-	Y

Figure 1.47   Very small numbers

UNIT 24
The Googol is Much Older Than Google

We are all familiar with the ubiquitous search engine Google and the phrase that has entered our vocabulary: "Google it". However, the word Google was coined as a mathematical term over a hundred years ago and was spelled *Googol*. In the year 1940, it was popularly introduced in the book *Mathematics and the Imagination*[19] written by Edward Kasner (1878–1955) and James Newman (1907–1966). In the book the appellation "googol" is ascribed to the enormous quantity 10 to the 100th power, or 1 followed by a hundred zeros. In 1920, Edward Kasner was working with very large numbers and on a walk along the New Jersey Palisades with his young nephew, Milton Sirotta (1911–1981), Kasner asked him for a name to call the number

---

[19] Simon and Schuster, New York: 1940.

10 to the 100th power. Nine-year-old Milton uttered something like "googoo" which became "googol", and satisfied Dr. Kasner's need for a fanciful label to pique young people's interest.

The name is in common use today in mathematics, however formal names for large numbers begin with a million: 1,000,000, which is 1 followed by 2 groups of 3 zeros. Add 1 more group of 3 zeros and you have a billion: 1,000,000,000, or 10^9, which employs the prefix bi, meaning two. Following this questionable logic, we have trillion: 1,000,000,000,000, quadrillion: 1,000,000,000,000,000, quintillion: 1,000,000,000,000,000,000, etc. which respectively have 4, 5 and 6 groups of three zeros. Technically then, it turns out that a googol is actually 10 dotrigintillions, where one dotrigintillion is 10^{99} and $10 \times 10^{99} = 10^{100}$. Using this terminology, the largest named number is centillion, which is 10^{303}. Dr. Kasner, at the same time, came up with the name *googolplex*, which is almost infinite, but not quite, and is one with a googol zeros after it or 10^{googol}. A googolplex has so many zeros that a piece of paper long enough to reach from here to the moon could not accommodate all of them.

As a further note, two English mathematicians, John Conway (1937–2020) and Richard Guy (1916–2020), have suggested that *N-plex* be used as a name for 10^N. This gives rise to the name *googolplexplex* for $10^{googolplex} = 10^{10^{10^{100}}}$. They also proposed that *N-minex* be used as a name for 10^{-N}, giving rise to the name *googolminex* for the reciprocal of a googolplex. Unlike Googol these names have not achieved any prominence in mathematics.

Origin of the name "Google"

According to David Skoller from Stanford University[20] where the name Google was born, Larry Page and Sergey Brin, the founders of Google, originally used the term "BackRub" for their search engine in 1996, inspired by its analysis of the web's "back links." As the internet

---

[20] https://www.webcitation.org/68ubHzYs7?url=http://graphics.stanford.edu/~dk/google_name_origin.html

technology rapidly improved, Page, along with his officemates in 1997, then brainstormed possible new names for their search engine. Considering the immense amount of data to be analyzed, a partner Sean Anderson, suggested the word "googolplex." Page responded verbally with the shortened form, "googol" and Brin checked the Internet domain name registry database to see if this name was still available for registration and use. However, he made the mistake of searching for the name spelled as "google.com," which he found to be available. Page responded favorably and moved quickly to register the name "google.com" for himself and Brin on September 15, 1997. As a further inspiration, the name Googleplex was then applied to their Corporate Headquarters.

UNIT 25

Armstrong or Narcissistic Numbers

It is always curious as to how peculiar number relationships become named. In the mid-1960s a mathematics professor at the University of Rochester, Michael F Armstrong, discovered an unusual number pattern, which he gave to his students to further investigate using early computer technology. Armstrong believed that the name got attached to these numbers in a February 23, 1988, column by Tim Hartnell in the newspaper *The Australian*. He claims this was his moment of fame. These *Armstrong* numbers, which we will investigate here, are also frequently referred to as *narcissistic* numbers. However, it must be revealed that the British mathematician Godfrey H. Hardy (1877–1947) referred to these numbers in his 1940 book *A Mathematician's Apology*.

Let us now investigate these narcissistic or Armstrong numbers. An Armstrong number is one which is equal to the sum of its digits, where each digit is taken to the power equal to the number of digits in the original number. For example, the three-digit number $153 = 1^3 + 5^3 + 3^3$. Since 153 is a 3-digit number each of the digits is taken to the third power and the sum is equal to 153. Strangely enough, while there are no 2-digit Armstrong Numbers, there are only four 3-digit Armstrong numbers as we show in the table in Figure 1.48.

n	$100 \leq n \leq 999$	
153	$1^3 + 5^3 + 3^3 = 1 + 125 + 27$	$= 153$
370	$3^3 + 7^3 + 0^3 = 9 + 343 + 0$	$= 370$
371	$3^3 + 7^3 + 1^3 = 9 + 343 + 1$	$= 371$
407	$4^3 + 0^3 + 7^3 = 64 + 0 + 343$	$= 407$

Figure 1.48

Now if we take this one step further, we see there are only three 4-digit Armstrong numbers that exist as we show in Figure 1.49.

n	$1,000 \leq n \leq 9,999$	
1,634	$1^4 + 6^4 + 3^4 + 4^4 = 1 + 1,296 + 81 + 256$	$= 1,634$
8,208	$8^4 + 2^4 + 0^4 + 8^4 = 4,096 + 16 + 0 + 4,096$	$= 8,208$
9,474	$9^4 + 4^4 + 7^4 + 4^4 = 6,561 + 256 + 2,401 + 256$	$= 9,474$

Figure 1.49

Among the 5-digit numbers, there are only three Armstrong numbers which are shown in Figure 1.50.

n	$10,000 \leq n \leq 99,999$	
54,748	$5^5 + 4^5 + 7^5 + 4^5 + 8^5 = 3,125 + 1,024 + 16,807 + 1,024 + 32,768$	$= 54,748$
92,727	$9^5 + 2^5 + 7^5 + 2^5 + 7^5 = 59,049 + 32 + 16,807 + 32 + 16,807$	$= 92,727$
93,084	$9^5 + 3^5 + 0^5 + 8^5 + 4^5 = 59,049 + 243 + 0 + 32,768 + 1,024$	$= 93,084$

Figure 1.50

Moving along we find the only Armstrong number among the 6-digit numbers to be 548,834, as shown here: $5^6 + 4^6 + 8^6 + 8^6 + 3^6 + 4^6 = 15,625 + 4,096 + 262,144 + 262,144 + 729 + 4,096 = 548,834$.

For the motivated reader, we offer the following table (Figure 1.51) that lists the Armstrong numbers of $7 - 10$ digits.

Number of digits	Armstrong-Numbers
7	1,741,725; 4,210,818; 9,800,817; 9,926,315
8	24,678,050; 24,678,051; 88,593,477
9	146,511,208; 472,335,975; 534,494,836; 912,985,153
10	4,679,307,774

Figure 1.51

Just to show what can be done with today's computers; we can come up with a number that is 39 digits long: 115,132,219,018,763, 992,565,095,597,973,971,522,401 and is equal to the sum of its digits, each of which is taken to the 39th power, thus making it a very large Armstrong number. Furthermore, it is the largest Armstrong number found to date. Here is its expansion:

$$1^{39} + 1^{39} + 5^{39} + 1^{39} + 3^{39} + 2^{39} + 2^{39} + 1^{39} + 9^{39} + 0^{39} + 1^{39} + 8^{39} + 7^{39} +$$
$$6^{39} + 3^{39} + 9^{39} + 9^{39} + 2^{39} + 5^{39} + 6^{39} + 5^{39} + 0^{39} + 9^{39} + 5^{39} + 5^{39} + 9^{39} +$$
$$7^{39} + 9^{39} + 7^{39} + 3^{39} + 9^{39} + 7^{39} + 1^{39} + 5^{39} + 2^{39} + 2^{39} + 4^{39} + 0^{39} + 1^{39} =$$
$$115,132,219,018,763,992,565,095,597,973,971,522,401$$
$$\approx 1.151322190 \times 10^{38}.$$ For more Armstrong numbers see Figure 1.52.

There are no Armstrong numbers for $k = 2$, 12, 13, 15, 18, 22, 26, 28, 30 and 36. In fact, there are only 89 Armstrong numbers in the decimal system. The following is a list of Armstrong Numbers that are *consecutive*.

$k = 3$:   370; 371

$k = 8$:   24,678,050; 24,678,051

$k = 11$:   32,164,049,650; 32,164,049,651

$k = 16$:   4,338,281,769,391,370; 4,338,281,769,391,371

$k = 25$:   3,706,907,995,955,475,988,644,380; 3,706,907,995,955,47
5,988,644,381

$k = 29$:   19,008,174,136,254,279,995,012,734,740 19,008,174,136,2
54,279,995,012,734,741

$k = 33$:   186,709,961,001,538,790,100,634,132,976,990; 186,709,96
1,001,538,790,100,634,132,976,991

$k = 39$:   115,132,219,018,763,992,565,095,597,973,971,522,400;
115,132,219,018,763,992,565,095,597,973,971,522,401

Incidentally, our first Armstrong Number, 153, has some other amazing properties:

It is also a triangular number[21] as follows:

---

[21] A triangular number is one that can represent the number of dots arranged in the shape of an equilateral triangle. These numbers are presented in greater depth on page 67.

k	Armstrong-Numbers of k-digit lengths
1	0; 1; 2; 3; 4; 5; 6; 7; 8; 9
3	153; 370; 371; 407
4	1634; 8208; 9474
5	54748; 92727; 93084
6	548834
7	1741725; 4210818; 9800817; 9926315
8	24678050; 24678051; 88593477
9	146511208; 472335975; 534494836; 912985153
10	4679307774
11	32164049650; 32164049651; 40028394225; 42678290603; 44708635679; 49388550606; 82693916578; 94204591914
14	28116440335967
16	4338281769391370; 4338281769391371
17	21897142587612075; 35641594208964132; 35875699062250035
19	1517841543307505039; 3289582984443187032; 4498128791164624869; 4929273885928088826
20	63105425988599693916
21	128468643043731391252; 449177399146038697307
23	21887696841122916288858; 27879694893054074471405; 27907865009977052567814; 28361281321319229463398; 35452590104031691935943
24	174088005938065293023722; 188451485447897896036875; 239313664430041569350093
25	1550475334214501539088894; 1553242162893771850669378; 3706907995955475988644380; 3706907995955475988644381; 4422095118095899619457938
27	121204998563613372405438066; 121270696006801314328439376; 128851796696487777842012787; 174650464499531377631639254; 177265453171792792366489765
29	14607640612971980372614873089; 19008174136254279995012734740; 19008174136254279995012734741; 23866716435523975980390369295
31	1145037275765491025924292050346; 1927890457142960697580636236639; 2309092682616190307509695338915
32	17333509977822493087251039627 72
33	186709961001538790100634132976990; 186709961001538790100634132976991
34	1122763285329372541592822900204593
35	12639369517103790328947807201478392; 12679937780272278566303885594196922
37	1219167219625434121569735803609966019
38	12815792078366059955099770545296129367
39	115132219018763992565095597973971522400; 115132219018763992565095597973971522401

Figure 1.52

$$1 + 2 + 3 + 4 + 5 + 6 + 7 + 8 + 9 + 10 + 11 + 12 + 13 + 14 + 15 + 16 + 17 = 153.$$

Furthermore, it is a number that can be expressed as the sum of consecutive factorials: $1! + 2! + 3! + 4! + 5! = 153$. Can you discover other properties of this ubiquitous number 153? Enjoy!

Just for "entertainment", here is another property of three numbers where the sum of the digits consecutively taken to the third power eventually brings us back to the original number as shown here with the number 55:

$$55: 5^3 + 5^3 = 250$$
$$250: 2^3 + 5^3 + 0^3 = 133$$
$$133: 1^3 + 3^3 + 3^3 = 55.$$

There are three other 3-digit numbers for which this pattern is true and we leave it to the motivated reader to verify it. The numbers are 136, 160, and 919.

UNIT 26

Friendly Numbers or Amicable Numbers

What could possibly make 2 numbers friendly or amicable? Mathematicians have decided that 2 numbers are to be considered "friendly" if the sum of the proper divisors[22] (or factors) of one number equals the second number *and* the sum of the proper divisors of the second number equals the first number. The smallest friendly pair of numbers is 220 and 284. To substantiate this, we take the divisors of **220** which are 1, 2, 4, 5, 10, 11, 20, 22, 44, 55, and 110, whose sum is $1 + 2 + 4 + 5 + 10 + 11 + 20 + 22 + 44 + 55 + 110 = $ **284**. Now we take the divisors of **284** which are 1, 2, 4, 71, and 142, and find their sum, which is $1 + 2 + 4 + 71 + 142 = $ **220**. This shows the two numbers can be considered *friendly numbers* or *amicable numbers*.

---

[22] Proper divisors are all the divisors or factors of the number except the number itself. For example, the proper divisors of 6 are 1, 2, and 3, but not 6.

The second pair of friendly numbers, 17,296 and 18,416, was discovered by the French mathematician Pierre de Fermat (1607–1665). We can substantiate the friendliness by identifying the prime divisors of each number as follows $17{,}296 = 2^4 \times 23 \times 47$, and $18{,}416 = 2^4 \times 1151$ and then finding the sum of all the divisors. Curiously, in recent years it has been previously discovered that these 2 numbers may have been identified as friendly numbers by the Moroccan mathematician Ibn al-Banna al-Marrakushi al-Azdi (1256–1321).

In order to substantiate the friendliness of these 2 numbers, we find the sum of the factors of 17,296, which is:

$$1 + 2 + 4 + 8 + 16 + 23 + 46 + 47 + 92 + 94 + 184 + 188 + 368 + 376 + 752 + 1081 + 2162 + 4324 + 8648 = 18416$$

We then find the sum of the factors of 18,416 which is:

$$1 + 2 + 4 + 8 + 16 + 1151 + 2302 + 4604 + 9208 = 17296$$

Hence, they are truly amicable numbers!

The French mathematician, René Descartes (1596–1650) found another pair of amicable numbers: 9,363,584, and 9,437,056. Later in 1747, the Swiss mathematician Leonhard Euler (1707–1783) discovered 60 pairs of amicable numbers, yet he seemed to have overlooked the second smallest pair, namely, 1,184 and 1,210, which was discovered in 1866 by the 16-year-old B. Nicolò I. Paganini. This can be seen as follows: The sum of the factors of 1,184 is $1 + 2 + 4 + 8 + 16 + 32 + 37 + 74 + 148 + 296 + 592 = 1210$. The sum of the factors of 1,210 is $1 + 2 + 5 + 10 + 11 + 22 + 55 + 110 + 121 + 242 + 605 = 1184$.

To date, we have identified over 363,000 pairs of amicable numbers, yet we do not know if there exists an infinite number of such pairs. The table in Figure 1.53 provides a list of the first 108 amicable numbers.

Going beyond this list, we will eventually stumble on an even larger pair of amicable numbers, such as 111,448,537,712 and 118,853,793,424.

	First Number	Second Number	Year of discovery
1	220	284	ca. 500 BCE-
2	1184	1210	1860
3	2620	2924	1747
4	5020	5564	1747
5	6232	6368	1747
6	10744	10856	1747
7	12285	14595	1939
8	17296	18416	ca. 1310/1636
9	63020	76084	1747
10	66928	66992	1747
11	67095	71145	1747
12	69615	87633	1747
13	79750	88730	1964
14	100485	124155	1747
15	122265	139815	1747
16	122368	123152	1941/42
17	141664	153176	1747
18	142310	168730	1747
19	171856	176336	1747
20	176272	180848	1747
21	185368	203432	1966
22	196724	202444	1747
23	280540	365084	1966
24	308620	389924	1747
25	319550	430402	1966
26	356408	399592	1921
27	437456	455344	1747
28	469028	486178	1966
29	503056	514736	1747
30	522405	525915	1747
31	600392	669688	1921
32	609928	686072	1747
33	624184	691256	1921
34	635624	712216	1921
35	643336	652664	1747
36	667964	783556	1966
37	726104	796696	1921
38	802725	863835	1966
39	879712	901424	1966
40	898216	980984	1747
41	947835	1125765	1946
42	998104	1043096	1966
43	1077890	1099390	1966
44	1154450	1189150	1957
45	1156870	1292570	1946
46	1175265	1438983	1747
47	1185376	1286744	1929
48	1280565	1340235	1747
49	1328470	1483850	1966
50	1358595	1486845	1747
51	1392368	1464592	1747
52	1466150	1747930	1966

Figure 1.53

	First Number	Second Number	Year of discovery
53	1468324	1749212	1967
54	1511930	1598470	1946
55	1669910	2062570	1966
56	1798875	1870245	1967
57	2082464	2090656	1747
58	2236570	2429030	1966
59	2652728	2941672	1921
60	2723792	2874064	1929
61	2728726	3077354	1966
62	2739704	2928136	1747
63	2802416	2947216	1747
64	2803580	3716164	1967
65	3276856	3721544	1747
66	3606850	3892670	1967
67	3786904	4300136	1747
68	3805264	4006736	1929
69	4238984	4314616	1967
70	4246130	4488910	1747
71	4259750	4445050	1966
72	4482765	5120595	1957
73	4532710	6135962	1957
74	4604776	5162744	1966
75	5123090	5504110	1966
76	5147032	5843048	1747
77	5232010	5799542	1967
78	5357625	5684679	1966
79	5385310	5812130	1967
80	5459176	5495264	1967
81	5726072	6369928	1921
82	5730615	6088905	1966
83	5864660	7489324	1967
84	6329416	6371384	1966
85	6377175	6680025	1966
86	6955216	7418864	1946
87	6993610	7158710	1957
88	7275532	7471508	1967
89	7288930	8221598	1966
90	7489112	7674088	1966
91	7577350	8493050	1966
92	7677248	7684672	1884
93	7800544	7916696	1929
94	7850512	8052488	1966
95	8262136	8369864	1966
96	8619765	9627915	1957
97	8666860	10638356	1966
98	8754130	10893230	1946
99	8826070	10043690	1967
100	9071685	9498555	1946
101	9199496	9592504	1929
102	9206925	10791795	1967
103	9339704	9892936	1966
104	9363584	9437056	ca. 1600/1638

Figure 1.53   (*Continued*)

	First Number	Second Number	Year of discovery
105	9478910	11049730	1967
106	9491625	10950615	1967
107	9660950	10025290	1966
108	9773505	11791935	1967

Figure 1.53 *(Continued)*

In order to pursue a search for additional amicable numbers, we can use the following method for finding such numbers:

Consider the following relationships:

$a = 3 \times 2^n - 1$

$b = 3 \times 2^{n-1} - 1$, and

$c = 3^2 \times 2^{2n-1} - 1$, where n is an integer ≥ 2.

If a, b and c are prime numbers, then (2^n) ab and (2^n) c are amicable numbers. (For n $\leq$ 200, only $n = 2, 4$, and 7 would give us a, b and c to be prime numbers.)

Inspecting the list of amicable numbers in Figure 1.53, we notice that each pair is either a pair of odd numbers or a pair of even numbers. To date, we do not know if there is a pair of amicable numbers, where one is odd, and one is even. We also do not know if any pair of amicable numbers is relatively prime (that is, the numbers have no common factor other than 1). These open questions contribute to our continued fascination with amicable numbers and provide constant challenges for future mathematicians.

UNIT 27

Happy and Unhappy Numbers

Curiosities in mathematics sometimes appear and their origin is often not quite clear. A mathematics professor at Leeds University, Reg Allenby, was approached by his daughter who encountered a curiosity in high school and brought it to his attention. It is believed that the name "happy numbers" originated at that point. What we have here is another rather strange curiosity in our decimal number system that leads us to partition our numbers into two categories. We say that a number is considered a *happy number*, if we take the sum of the squares of the digits to get a second number, and then take the sum of the squares of the digits of that second number to get a third number, and then take the sum of the squares of the digits of the third number, and then continue until you get to the number 1. The numbers, where this scheme is applied and do *not* eventually end with the number 1 are called *unhappy numbers*. These latter numbers will eventually end up in a loop — where they will continue through a cycle.

Let's consider one of the happy numbers, say 13, and follow this process of taking the sum of the squares of the digits continuously.

$$1^2 + 3^2 = 10$$
$$1^2 + 0^2 = 1$$

Let's consider a slightly longer path to the number 1, by using 19.

$$1^2 + 9^2 = 82$$
$$8^2 + 2^2 = 68$$
$$6^2 + 8^2 = 100$$
$$1^2 + 0^2 + 0^2 = 1$$

Following is a list of the happy numbers from 1 to 1000. You might want to try a few of these to test their "happiness".

1, 7, 10, 13, 19, 23, 28, 31, 32, 44, 49, 68, 70, 79, 82, 86, 91, 94, 97, 100, 103, 109, 129, 130, 133, 139, 167, 176, 188, 190, 192, 193, 203, 208, 219, 226, 230, 236, 239, 262, 263, 280, 291, 293, 301, 302, 310, 313, 319, 320, 326, 329, 331, 338, 356, 362, 365, 367, 368, 376, 379, 383, 386, 391, 392, 397, 404, 409, 440, 446, 464, 469, 478, 487, 490, 496, 536, 556, 563, 565, 566, 608, 617, 622, 623, 632, 635, 637, 638, 644, 649, 653, 655, 656, 665, 671, 673, 680, 683, 694, 700, 709, 716, 736, 739, 748, 761, 763, 784, 790, 793, 802, 806, 818, 820, 833, 836, 847, 860, 863, 874, 881, 888, 899, 901, 904, 907, 910, 912, 913, 921, 923, 931, 932, 937, 940, 946, 964, 970, 973, 989, 998, 1000.

The first few consecutive happy numbers (n; $n + 1$) are: (31, 32), (129, 130), (192, 193), (262, 263), (301, 302), (319, 320), (367, 368), (391, 392), ...

As you will see when you try to discover happy numbers, many will fall into similar paths, such as the numbers 19 and 91. Following are the unique combinations, that is, none of these numbers will take the same path as they get to the number 1. Naturally, the above list includes variations of these.

1, 7, 13, 19, 23, 28, 44, 49, 68, 79, 129, 133, 139, 167, 188, 226, 236, 239, 338, 356, 367, 368, 379, 446, 469, 478, 556, 566, 888, 899.

It appears that happy numbers shown above are without zeros, and with digits in increasing order.

An example of an unhappy number is 25. Notice how this procedure will lead us into a loop and never to reach the number 1.

$$2^2 + 5^2 = 29$$
$$2^2 + 9^2 = 85$$
$$8^2 + 5^2 = \underline{89}$$
$$8^2 + 9^2 = 145$$
$$1^2 + 4^2 + 5^2 = 42$$
$$4^2 + 2^2 = 20$$
$$2^2 + 0^2 = 4$$
$$4^2 = 16$$
$$1^2 + 6^2 = 37$$
$$3^2 + 7^2 = 58$$
$$5^2 + 8^2 = \underline{89}$$

This then begins another loop — repeating everything again with 89, which was reached earlier in this process.

Among the happy numbers there are those that are prime numbers as well — known as *happy prime numbers*. The happy prime numbers less than 500 are the following:

7, 13, 19, 23, 31, 79, 97, 103, 109, 139, 167, 193, 239, 263, 293, 313, 331, 367, 379, 383, 397, 409, 487.

To date, mathematicians have made some claims as to some special happy numbers. For example, the smallest happy number that exhibits all the digits 0 – 9 is 10,234,456,789. While the smallest happy number that has no zeros, and yet has all the other digits is 1,234,456,789. Then there is the smallest happy number that has no zeros and is also palindromic: 13,456,789,298,765,431. When we accept the inclusion of zeros, the smallest happy number with all the digits is 1,034,567,892,987,654,301. Thus far, the largest happy number where no digit is repeated is 986,543,210.

UNIT 28

Repunits, Numbers Consisting Only of 1's

When teachers at the secondary school level want to enrich their instructional program, they often mention some of the most unusual numbers, which from their appearance have gotten the name "repunits." These are numbers consisting of only the digit 1, such as the numbers 1, 11, 111, 1111, 11111, ... , and they harbor some rather unusual and curious properties. The name *repunits* was coined in 1964 by the American mathematician Albert H. Beiler (1909–1973) in his book *Recreations in the Theory of Numbers*. It simply means numbers consisting of the repetition of the units digit. These numbers can be expressed algebraically in the form $\frac{10^n-1}{9}$, where n represents the number of 1's in the repunits number.

One of the curiosities that is easily identified occurs when we take the square of a repunits number. The result is a palindromic number, which is a number that reads backwards and forwards the same as we can see in the table in Figure 1.54.

Number of 1's	Repunits number	Square of Repunits number
1	1	1
2	11	121
3	111	12321
4	1111	1234321
5	11111	123454321
6	111111	12345654321
7	1111111	1234567654321
8	11111111	123456787654321
9	111111111	12345678987654321

Figure 1.54

If one were to take this process further, the symmetric pattern would continue indefinitely.

Many of the repunits (numbers consisting of only the digit 1) are composite numbers, that is the product of various prime numbers. However, we notice that there are some repunits such as shown in Figure 1.55, where r_2 and r_{19} are prime numbers. The question then arises are there other such repunits which are prime numbers? The answer is yes. However, mathematicians have struggled with this question for years. For example, the German mathematician Gustav Jacob Jacobi (1804–1851) sought to determine if the repunit r_{11} is a prime number. Much to his disappointment, we find that with today's technology we can determine that it is not a prime number in less

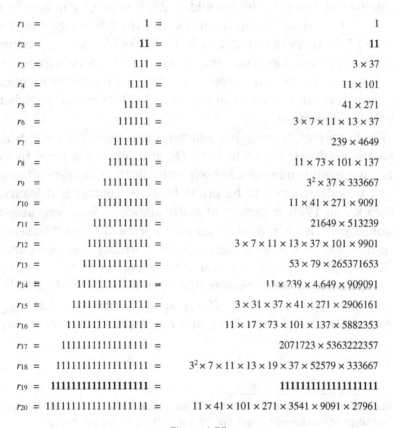

$$
\begin{aligned}
r_1 &= 1 = 1 \\
r_2 &= 11 = 11 \\
r_3 &= 111 = 3 \times 37 \\
r_4 &= 1111 = 11 \times 101 \\
r_5 &= 11111 = 41 \times 271 \\
r_6 &= 111111 = 3 \times 7 \times 11 \times 13 \times 37 \\
r_7 &= 1111111 = 239 \times 4649 \\
r_8 &= 11111111 = 11 \times 73 \times 101 \times 137 \\
r_9 &= 111111111 = 3^2 \times 37 \times 333667 \\
r_{10} &= 1111111111 = 11 \times 41 \times 271 \times 9091 \\
r_{11} &= 11111111111 = 21649 \times 513239 \\
r_{12} &= 111111111111 = 3 \times 7 \times 11 \times 13 \times 37 \times 101 \times 9901 \\
r_{13} &= 1111111111111 = 53 \times 79 \times 265371653 \\
r_{14} &= 11111111111111 = 11 \times 239 \times 4,649 \times 909091 \\
r_{15} &= 111111111111111 = 3 \times 31 \times 37 \times 41 \times 271 \times 2906161 \\
r_{16} &= 1111111111111111 = 11 \times 17 \times 73 \times 101 \times 137 \times 5882353 \\
r_{17} &= 11111111111111111 = 2071723 \times 5363222357 \\
r_{18} &= 111111111111111111 = 3^2 \times 7 \times 11 \times 13 \times 19 \times 37 \times 52579 \times 333667 \\
r_{19} &= \mathbf{1111111111111111111} = \mathbf{1111111111111111111} \\
r_{20} &= 11111111111111111111 = 11 \times 41 \times 101 \times 271 \times 3541 \times 9091 \times 27961
\end{aligned}
$$

Figure 1.55

than a second's time. Factoring repunit numbers is often very difficult; however, with the aid of a computer we find that, for example, the repunit r_{71} is factorable as follows, and therefore, it is not a prime number.

r_{71} = 11111111111111111111111111111111111111111111111111
1111111111111111111111
= (241573142393627673576957439049)
× (45994811347886846310221728895223034301839).

As we mentioned earlier, the next repunit number after 11 that is a prime has 19 digits as follows: 1,111,111,111,111,111,111. Were we to search for the next larger repunit number that is a prime number, the number of digits required would be 23. Following that, in successive order, the prime repunit numbers have the following number of digits: 317, 1031, 49081, 86453, 109297, 270343. There may be more prime repunit numbers, but they have not yet been discovered. Clearly the remainder of the repunit numbers are composite numbers and we offer the first few here in Figure 1.55, where we factored them into their prime factors.

The history of determining which repunit numbers were prime has a curious development. In 1918, Oscar Hoppe established that r_2 and r_{19} were prime numbers. Subsequently, in 1929, the repunit number r_{23} was determined to be prime by D. H. Lehmer and Maurice Kraitchik.[23] In 1970, a student of mathematics, E. Seah, was able to demonstrate that the repunit r_{317} is also a prime number. The search for prime numbers among the repunits continues. For example, in 1985 the repunit r_{1031} was discovered to be a prime by H. C. Williams and H. Dubner. Further repunits that have been since identified as primes are: r_{49081} (H. Dubner, 1999), r_{86453} (L. Baxter, 2000), r_{109297} (P. Bourdelais, H.Dubner, 2007), and r_{270343} (M. Voznyy, A. Budnyy,

---

[23] Francis, Richard L.; "Mathematical Haystacks: Another Look at Repunit Numbers" in *The College Mathematics Journal*, Vol. 19, No. 3. (May, 1988), pp. 240–246.

2007) [24]. It is believed today that there are an infinite number of repunits which are prime numbers.

For example, there are differences of squares which yield repunit numbers as we see in Figure 1.56.

Another curious pattern can be seen in Figure 1.57, by considering the second, fourth, and sixth entries from Figure 1.56, we notice an additional pattern generating numbers of particular interest each time an additional 5 and 4 is tagged onto the front of the numbers, respectively. If we continue this pattern, we notice the surprising result that evolves, which is shown in Figure 1.57.

$$1^2 - 0^2 = 1$$
$$6^2 - 5^2 = 11$$
$$20^2 - 17^2 = 111$$
$$56^2 - 45^2 = 1,111$$
$$156^2 - 115^2 = 11,111$$
$$556^2 - 445^2 = 111,111$$
$$344^2 - 85^2 = 111,111$$
$$356^2 - 125^2 = 111,111$$

Figure 1.56

$$6^2 - 5^2 = 11$$
$$56^2 - 45^2 = 1,111$$
$$556^2 - 445^2 = 111,111$$
$$5556^2 - 4445^2 = 11,111,111$$
$$55556^2 - 44445^2 = 1,111,111,111$$
$$555556^2 - 444445^2 = 111,111,111,111$$
$$5555556^2 - 4444445^2 = 11,111,111,111,111$$
$$55555556^2 - 44444445^2 = 1,111,111,111,111,111$$
$$555555556^2 - 444444445^2 = 111,111,111,111,111,111$$
$$]$$

$$55555555555555556^2 - 44444444444444445^2 =$$

$$1,111,111,111,111,111,111,111,111,111,111,111$$

Figure 1.57

---

[24] http://mathworld.wolfram.com/Repunit.html

Then there are more curious numbers; those that are generated by the difference of squares and are multiples of repunit numbers, as shown in Figure 1.58.

In Figure 1.59, we notice another such pattern of numbers, which should be admired.

Our further investigation of repunit numbers brings us to an interesting pattern; one where we divide 111,111,111 by 9, which gives us the number 12,345,679. Notice we have the digits in numerical order, but we are missing the number 8. Yet, when we consider the following pattern (Figure 1.60), the 8 is once again included in generating repunit numbers.

Surprisingly, this pattern can also be extended to the following: 123,456,789 × 9 + 10 = 1,111,111,111. As you can see, repunit

$$7^2 - 4^2 = 33 = 3 \cdot 11$$
$$67^2 - 34^2 = 3,333 = 3 \cdot 1,111$$
$$667^2 - 334^2 = 333,333 = 3 \cdot 111,111$$
$$6667^2 - 3334^2 = 33,333,333 = 3 \cdot 11,111,111$$
$$66667^2 - 33334^2 = 3,333,333,333 = 3 \cdot 1,111,111,111$$

Figure 1.58

$$8^2 - 3^2 = 55 = 5 \cdot 11$$
$$78^2 - 23^2 = 5555 = 5 \cdot 1,111$$
$$778^2 - 223^2 = 555,555 = 5 \cdot 111,111$$
$$7778^2 - 2223^2 = 55,555,555 = 5 \cdot 11,111,111$$
$$77778^2 - 22223^2 = 5,555,555,555 = 5 \cdot 1,111,111,111$$

Figure 1.59

$$
\begin{array}{rcrcrcl}
0 & \times & 9 & + & 1 & = & 1 \\
1 & \times & 9 & + & 2 & = & 11 \\
12 & \times & 9 & + & 3 & = & 111 \\
123 & \times & 9 & + & 4 & = & 1,111 \\
1,234 & \times & 9 & + & 5 & = & 11,111 \\
12,345 & \times & 9 & + & 6 & = & 111,111 \\
123,456 & \times & 9 & + & 7 & = & 1,111,111 \\
1,234,567 & \times & 9 & + & 8 & = & 11,111,111 \\
12,345,678 & \times & 9 & + & 9 & = & 111,111,111 \\
\end{array}
$$

Figure 1.60

numbers seem to generate some rather interesting relationships and patterns, and the development of their history is an ongoing challenge to mathematicians.

UNIT 29

Palindromic Numbers

A palindrome, as you are probably aware, is a word or a sentence that when the letters are written backwards, they are in the same order. In mathematics, a palindromic number is one that is the same when the digits are reversed. These numbers that we encounter on a regular basis, sometimes cause us to marvel at them. They could be numbers with all the same digits, such as the number 555555, or they might be numbers that read the same in both directions such as the number 1234321.

The word itself is from the Greek *palindromus*, which means to run back again, which is what happens when you read the phrase or number backwards. Some interesting palindromes are shown in Figure 1.61. The Greek poet, Sotades, who lived in Egypt in about 276 BCE during the reign of Ptolemy II, wrote a palindrome with which he insulted the king. His gruesome punishment had him sealed in a box

A
EVE
RADAR
REVIVER
ROTATOR
LEPERS REPEL
MADAM I'M ADAM
STEP NOT ON PETS
DO GEESE SEE GOD
PULL UP IF I PULL UP
NO LEMONS, NO MELON
DENNIS AND EDNA SINNED
ABLE WAS I ERE I SAW ELBA
A MAN, A PLAN, A CANAL, PANAMA
A SANTA LIVED AS A DEVIL AT NASA
SUMS ARE NOT SET AS A TEST ON ERASMUS
ON A CLOVER, IF ALIVE, ERUPTS A VAST, PURE EVIL, A FIRE VOLCANO

Figure 1.61   Palindromes

and cast out to sea. Palindromes are frequently found written in Egyptian hieroglyphics and in Latin texts.

There is a well-known Latin palindromic sentence that stems from the 2nd century CE and has an additional amazing property. It reads: "sator arepo tenet opera rotas", which commonly translates to "Arepo the sower holds the wheels at work" (see Figure 1.62). The so-called Templar magic square places these letters in a five-by-five square arrangement. Now you can read the sentence down, up, from the left and from the right. This is quite astonishing! The Templar magic square has been found in excavations of the Roman city of Pompeii, which had been buried in the ashes from the 79 CE eruption of Mount Vesuvius. In medieval times, people have attributed magical properties to it and used it as a spell to protect against witchcraft.

Our concern here is to investigate the curiosities involved with *palindromic numbers*, which read the same in both directions: left to right, or right to left. A simple example which generates palindromic numbers is the first four powers of 11:

Figure 1.62   Templar magic square

$$11^0 = 1$$
$$11^1 = 11$$
$$11^2 = 121$$
$$11^3 = 1331$$
$$11^4 = 14641$$

Before we leave that lovely number 11, we should note that all palindromes with an even number of digits are divisible by 11. Technically all single-digit numbers are also palindromes. Taking this a step further, we also know that there are 9 palindromic numbers comprised of two digits, as the following: 11, 22, 33, 44, 55, 66, 77, 88, 99. There are 90 palindromic numbers with three digits, as follows: 101, 111, 121, 131, 141, 151, 161, 171, 181, 191, ..., 909, 919, 929, 939, 949, 959, 969, 979, 989, 999. Furthermore, there are also 90 palindromic numbers comprised with four digits: 1001, 1111, 1221, 1331, 1441, 1551, 1661, 1771, 1881, 1991, ..., 9009, 9119, 9229, 9339, 9449, 9559, 9669, 9779, 9889, 9999. Using this scheme, we can determine the number of palindromic numbers of other multi-digit numbers.

Dates sometimes provide us with a palindromic pattern, such as February 2, 2020, or 02-02-2020 is a palindromic date. On a more serious (mathematical) note, it is interesting to see how a palindromic number can be generated from most randomly selected numbers. All you need to do is to continually add the number to its reversal (that is, the number written in the reverse order of digits) until you arrive at a palindrome. For example, starting with the number 25 the sum 25 + 52 = 77 is a palindromic number. Or it might take two steps, such as with the starting number 76. The two successive sums 76 + 67 = 143, and 143 + 341 = 484, leads us to a palindromic number. Or it might take three steps, such as with the starting number 86: 86 + 68 = 154, 154 + 451 = 605, and 605 + 506 = 1111.

If we use 97 as the starting number, we will require 6 steps to reach a palindromic number; while the starting number 89 will require 24 steps to reach a palindromic number, as we show below.

1. $89 + 98 = 187$
2. $187 + 781 = 968$
3. $968 + 869 = 1837$
4. $1837 + 7381 = 9218$
5. $9218 + 8129 = 17347$
6. $17347 + 74371 = 91718$
7. $91718 + 81719 = 173437$
8. $173437 + 734371 = 907808$
9. $907808 + 808709 = 1716517$
10. $1716517 + 7156171 = 8872688$
11. $8872688 + 8862788 = 17735476$
12. $17735476 + 67453771 = 85189247$
13. $85189247 + 74298158 = 159487405$
14. $159487405 + 504784951 = 664272356$
15. $664272356 + 653272466 = 1317544822$
16. $1317544822 + 2284457131 = 3602001953$
17. $3602001953 + 3591002063 = 7193004016$
18. $7193004016 + 6104003917 = 13297007933$
19. $13297007933 + 33970079231 = 47267087164$
20. $47267087164 + 46178076274 = 93445163438$
21. $93445163438 + 83436154439 = 176881317877$
22. $176881317877 + 778713188671 = 955594506548$
23. $955594506548 + 845605495559 = 1801200002107$
24. $1801200002107 + 7012000021081 = 8813200023188$

There are other peculiarities involving palindromic numbers such as the fact that there are twelve numbers less than 1000 for which the reverse-sum sequence leads to the palindromic number 8813200023188. One of these numbers 484, is itself a palindromic number. A more startling fact is that the number 196 has not yet been shown to produce a palindromic number — even with over three million reversal additions. We still do not know if this one will ever reach a palindromic number. If you were to apply this procedure to 196, you would reach the number 227574622 at the 16th addition, a number you would also reach at the 15th step of the attempt to get a palindromic number from the starting number 788. This would then tell you that applying the procedure to the number 788 has also never been shown to reach a palindromic number. As a matter of fact, among the first 100,000 natural numbers, there are 5,996 numbers for which we have not yet been able to show that the procedure of reversal additions will lead to a palindromic number. Some of these unsuccessful starting numbers are: 196, 691, 788, 887, 1675, 5761, 6347, and 7436.

Using this procedure of reverse-and-add, we find that some numbers yield the same palindromic numbers in the same number of steps, such as 554, 752, and 653, which all produce the palindromic number 11011 in three steps. In general, all integers in which the corresponding digit pairs symmetric to the middle 5 have the same sum will produce the same palindromic number in the same number of steps. However, there are other integers that produce the same palindromic number, yet in a different number of steps, such as the number 198, which with repeated reversals and additions, will reach the palindromic number 79497 in five steps, while the number 7299 will reach this number in two steps.

For a two-digit number ab with digits $a \neq b$, the sum $a + b$ of its digits determines the number of steps needed to produce a palindrome. Clearly, if the sum of the digits is less than 10, then only one step will be required to reach a palindrome, for example, $25 + 52 = 77$. If the sum of the digits is 10, then $ab + ba = 110$, and $110 + 011 = 121$, and two steps will be required to reach the palindrome.

Earlier in this chapter (page 90), we introduced numbers that consist entirely of 1's and are called *repunits*. All the repunit numbers with fewer than ten 1's, when squared, yield palindromic numbers. For example, $1111^2 = 1234321$.

There are also some palindromic numbers that, when cubed, yield again palindromic numbers.

To this class belong all numbers of the form $n = 10^k + 1$, for $k = 1$, 2, 3, ... When n is cubed, it yields a palindromic number, which has $k - 1$ zeros between each consecutive pair of 1, 3, 3, 1.

$k = 1, n = 11$: $11^3 = 1331$
$k = 2, n = 101$: $101^3 = 1030301$
$k = 3, n = 1001$: $1001^3 = 1003003001$

We can continue to generalize and get some interesting patterns such as when n consists of three 1's and any even number of zeros symmetrically placed between the end 1's when cubed will give us a palindromic number, such as

$$111^3 = 1367631$$
$$10101^3 = 1030607060301$$
$$1001001^3 = 1003006007006003001$$
$$100010001^3 = 1000300060007000600030001$$

Taking this even a step further, we find that when n consists of four 1's and zeros in a palindromic arrangement, where the places between the 1's do not have same number of zeros, then n^3 will also be a palindromic number, as we can see with the following examples:

$$11011^3 = 1334996994331$$
$$10100101^3 = 1030331909339091330301$$

However, when the same number of zeros appears between the ones then the cube of the number will not result in a palindrome, as in the following example: $1010101^3 = 1030610121210060301$. As a matter of fact, the number 2201 is the only non-palindromic number, which is less than 280,000,000,000,000, and, when cubed, yields a palindromic number: $2201^3 = 10662526601$.

However, just for amusement consider the following pattern with palindromic numbers:

$$12321 = \frac{333 \times 333}{1+2+3+2+1}$$

$$1234321 = \frac{4444 \times 4444}{1+2+3+4+3+2+1}$$

$$123454321 = \frac{55555 \times 55555}{1+2+3+4+5+4+3+2+1}$$

$$12345654321 = \frac{666666 \times 666666}{1+2+3+4+5+6+5+4+3+2+1}$$

and so on. An ambitious reader may search for other patterns involving palindromic numbers.

UNIT 30

The Ubiquitous Number 1089

We don't have much history to share about how the number 1089 became popular. However, the amazing characteristic of this number will allow us to show the power of algebraic thinking. We encourage the reader to join us as we select any three-digit number, where the unit and hundreds digit are not the same. Now follow along as we demonstrate the process with our arbitrarily selected number 218. In bold print are the instructions and our illustrated number is also bold.

Choose any three-digit number (where the unit and hundreds digit are not the same).

> We demonstrate by arbitrarily selecting: **218**

Reverse the digits of this number you have selected.

> We reverse the digits of 218 to get: **812**

Subtract the two numbers (naturally, the larger minus the smaller)

> Our calculated difference is: **812 − 218 = 594**

Once again, reverse the digits of this difference.

> Reversing the digits of 594 we get the number: **495**

Now, add your last two numbers.

> We then add the last two numbers to get: 594 + 495 = **1089**

Everyone's result should be the same as ours, even though each person's starting number was different. You may be astonished with the result. Regardless of which number was selected at the beginning, the same result was reached by all, namely, 1089. How does this happen? Is this a "freak property" of this number? Did we do something illegitimate in our calculations? Can we assume that any number we chose would lead us to 1089? How can we be sure? Well, we could try all possible three-digit numbers to see if the number 1089 will appear at the end. That would be tedious and not particularly elegant.

Following is a possible procedure to better understand what was going on during our calculation. We shall represent the arbitrarily selected three-digit number as **htu**, where h represents the hundreds digit, t represents the tens digit, and u represents the units digit. The value of the number is then: $100h + 10t + u$. Let $h > u$, which would be the case either in the number you selected or the reverse of it. Then in the subtraction, $u - h < 0$; therefore, to enable subtraction, take 1 from the tens place of *htu* (the minuend) making the units place $10 + u$. Since the tens digits of the two numbers to be subtracted are equal, and 1 was taken from the tens digit of the minuend, the value of this digit is now $10(t - 1)$. It is then necessary to reduce the hundreds digit of the minuend to $h - 1$, because of the 1 that was taken away to enable subtraction in the tens place. The value of the tens digit is now $10(t - 1) + 100 = 10(t + 9)$.

We can now do the first subtraction:

$$
\begin{array}{lll}
100(h-1) & +10(t+9) & +(u+10) \\
100u & +10t & +h \\
\hline
100(h-u-1) & +10(9) & +u-h+10
\end{array}
$$

Reversing the digits of this difference gives us:

$$100(u - h + 10) + 10(9) + (h - u - 1)$$

Now adding these last two expressions gives us:

$$[100(h - u - 1) + 10(9) + (u - h + 10)] + [100(u - h + 10) + 10(9) + (h - u - 1)] = 100(9) + 10(18) + (10 - 1) = 1089$$

By this exercise, we can appreciate how algebra allows one to inspect the arithmetic process regardless of the number and see it as a process beyond a mechanical task required in a mathematics curriculum.

Here is another opportunity associated with a beautiful pattern in the number **1089**. Look at the first ten multiples of 1089:

$$1089 \times 1 = 1089$$

$$1089 \times 2 = 2178$$

$$1089 \times 3 = 3267$$

$$1089 \times 4 = 4356$$

$$1089 \times 5 = 5445$$

$$1089 \times 6 = 6534$$

$$1089 \times 7 = 7623$$

$$1089 \times 8 = 8712$$

$$1089 \times 9 = 9801$$

The pattern among the products should be easily recognizable. Notice that the first and ninth products are reverses of one another. The second and the eighth are also reverses of one another. And so, the pattern continues until the fifth product is the reverse of itself, known as a palindromic number. There is more to admire with the number 1089 since it is equal to 33^2. Furthermore, it is the smallest 4-digit number whose reversal is related to the original number as $\frac{9801}{9} = 1089$. The next such number with this property is the number obtained by doubling 1089 to get 2178, where $\frac{8712}{4} = 2178$. These phenomena can and should motivate readers to search for other such curiosities.

UNIT 31

Kaprekar Numbers

Some numbers have unusual peculiarities. Sometimes these peculiarities can be understood and justified through an algebraic representation, while at other times a peculiarity is simply a quirk of the base-10 number system. In any case, these numbers provide us with some rather entertaining amusements that ought to motivate us to look for other such peculiarities or oddities.

Consider, for example, the number 297. When we take the square of that number, we get $297^2 = 88{,}209$, which, if we were to split it into two numbers, strangely enough, the sum of the two numbers results in the original number: $88 + 209 = 297$. Such a number is called a *Kaprekar number*, named after the Indian mathematician Dattaraya Ramchandra Kaprekar (1905–1986) who discovered such numbers. Figure 1.63 offers some more examples.

Kaprekar number	Square of the number			Decomposition of the number
1	1^2	=	1	1 = 1
9	9^2	=	81	8 + 1 = 9
45	45^2	=	2,025	20 + 25 = 45
55	55^2	=	3,025	30 + 25 = 55
99	99^2	=	9,801	98 + 01 = 99
297	297^2	=	88,209	88 + 209 = 297
703	703^2	=	494,209	494 + 209 =703
999	999^2	=	998,001	998 + 001 = 999
2,223	$2{,}223^2$	=	4,941,729	494 + 1,729 = 2,223
2,728	2728^2	=	7,441,984	744 + 1,984 = 2,728
4,879	$4{,}879^2$	=	23,804,641	238 + 04,641 = 4,879
4,950	$4{,}950^2$	=	24,502,500	2,450 + 2,500 = 4,950
5,050	$5{,}050^2$	=	25,502,500	2,550 + 2,500 = 5,050
5,292	$5{,}292^2$	=	28,005,264	28 + 005,264 = 5,292
7,272	$7{,}272^2$	=	52,881,984	5,288 + 1,984 = 7,272
7,777	$7{,}777^2$	=	60,481,729	6,048 + 1,729 = 7,777
9,999	999^2	=	99,980,001	9,998 + 0,001 = 9,999
17,344	$17{,}344^2$	=	300,814,336	3,008 + 14,336 = 17,344
22,222	$22{,}222^2$	=	493,817,284	4,938 + 17,284 = 22,222
142,857	$142{,}857^2$	=	20,408,122,449	20,408 + 122,449 = 142,857

Figure 1.63   Kaprekar numbers

Some larger Kaprekar numbers are 38962; 77778; 82656; 95121; 99999; ...;499500, 500500, 533170, 538461; 857143; .... An ambitious reader may want to verify that these are also legitimate Kaprekar numbers.

There are also further variations, such as the number 45, which we would consider a *Kaprekar triple*, since it behaves as follows: $45^3 = 91,125 = 9 + 11 + 25 = 45$. Other Kaprekar triples are: 1, 8, 10, 297, and 2322. Curiously enough the number 297, which we previously demonstrated as a Kaprekar number is also a Kaprekar triple, since $297^3 = 26,198,073$, and $26 + 198 + 073 = 297$. Readers may now be motivated to find other Kaprekar triples.

The Kaprekar Constant

While we are dealing with Kaprekar numbers, it would be appropriate to extend this to a very special number referred to as a Kaprekar constant. An oddity that is apparently a quirk of the base-10 number system is this *Kaprekar constant*, which is the number 6,174. This number arises when one takes any four-digit number (with not all digits the same) and forms the largest and the smallest number from these digits, and then subtracts these two newly-formed numbers. Continuously repeating this process with the resulting differences will eventually result in the number 6,174. When the number 6,174 is reached and the process is continued — that is, creating the largest and the smallest number, and then taking their difference (7641 − 1467 = 6174), we will always get back to 6,174. This is called the *Kaprekar constant*. To demonstrate this with an example, we will carry out this process with a randomly-selected number. As mentioned above, when choosing the number, avoid numbers with four identical digits, such as 3333. The difference between largest and smallest number must be at least 2. For numbers with less than four digits, you obtain four digits by padding the number with zeros on the left, such as 0012. For our example, we will choose the number 3,618:

- The largest number formed with these digits is: 8,631.
- The smallest number formed with these digits is: 1,368.
- The difference is: 7,263.
- The largest number formed with these digits is: 7,632.
- The smallest number formed with these digits is: 2,367.
- The difference is: 5,265
- The largest number formed with these digits is: 6,552.
- The smallest number formed with these digits is: 2,556.
- The difference is: 3,996.
- The largest number formed with these digits is: 9,963
- The smallest number formed with these digits is: 3,699
- The difference is: 6264
- The largest number formed with these digits is: 6642
- The smallest number formed with these digits is: 2466
- The difference is: 4176
- The largest number formed with these digits is: 7641
- The smallest number formed with these digits is: 1467
- The difference is: 6174
- The largest number formed with these digits is: 7641
- The smallest number formed with these digits is: 1467
- The difference is: 6174

In the case where you choose a 4-digit number, where the largest digit differs from the smallest digit by less than two, the result of this "Kaprekar process" will be zero. In all other cases, you will always end up with the number 6,174, which then gets you into an endless loop (i.e. continuously getting back to 6,174). It should never take more than seven subtractions. If it does, then there must have been a calculating error.

In case you are wondering if this technique can be done with 3-digit numbers (where all the digits are not the same), the answer is yes and eventually the number 495 would be reached.

Incidentally, another curious property of 6,174 is that it is divisible by the sum of its digits:

$$\frac{6174}{6+1+7+4} = \frac{6174}{18} = 343.$$

There is still another curiosity involving the sum of the digits, 18, of the number 6174, namely,

$18^1 + 18^2 + 18^3 = 18 + 324 + 5832 = 6174$. So, we can see that 6174 is truly a special number.

UNIT 32

The Transcendency of Transcendental Numbers

Transcendental is an impressive word for a number, but these numbers earn their handle and the reason for the name is explained below. At a young age, we are introduced to the most common transcendental number, π, which owes its existence to the perfect 2-dimensional figure, the circle. We learn that the circumference of a circle equals π times its diameter and that π is approximately $\frac{22}{7}$. It is certainly a ubiquitous number, but it is also a revered mathematical quantity. The number π is an irrational number, which means it cannot be accurately expressed as a fraction, and is an infinite decimal, possessing no pattern among its digits. One can almost trace the historical development of mathematics based on how well the number π was accurately calculated. In recent years, it has been computed to over 31.4 trillion of its digits by the Japanese computer scientist Emma Haruka Iwao (1984 –), who set the newest Guinness World Record for the most accurate value of π. This Google employee and her team calculated 31,415,926,535,897 digits of π – crushing a 2016 record.

In calculus another transcendental number appears, which is the number e, where e^x is the natural exponential growth function, which applies to early growth in plants and animals, but it is not as well-known as π. It is also called Euler's number and is approximately equal to 2.71828. One way to understand the significance of *e* is to see its application to compound interest.

These days money in a savings account is compounded every day and grows exponentially but it would not grow much faster even if it were compounded every second or every instant. This is shown by the formula for the instantaneous compounding of money: $A = Pe^{rn}$, where A is the final amount, P is the principal, r is the annual interest rate and n is the number of years. The transcendental number *e* shows us that there is a limit to exponential growth as it occurs more and more often.

The name transcendental, though, has to do with the role these numbers play in the solution of equations. A number is considered transcendental if it cannot be the root of a polynomial equation with rational coefficients, numbers which are either fractions or integers. For example, a second-degree quadratic equation of the form $ax^2 + bx + c = 0$ can never have π or e as a solution when a, b and c are rational numbers. No matter how high the degree of the polynomial equation, π or e can never be a solution.

All transcendental numbers, by nature of their definition, are irrational, and there is no lack of them. There are "more" transcendental numbers than there are rational numbers. What is meant by this, is that while the infinity of rational numbers is countable, the infinity of transcendental numbers is uncountable, which is a higher order of infinity.[25] Another example of a transcendental number is $\left(\sqrt{2}\right)^{\sqrt{2}}$, which is called the Hilbert number, after David Hilbert (1862–1943), a German mathematician and one of the most influential mathematicians of the 19th and early 20th centuries. Some other noteworthy examples are: e^π, sin A, where A is expressed in radians and $\log_b a$, where a and b are positive integers and not both powers of the same integer. Transcendental numbers, though intriguing to mathematicians, also transcend most people's everyday experiences, and thereby find little use in solving our practical problems.

---

[25]See The Infinity Concept and its Symbol, page 48.

Chapter 2

Arithmetic Curiosities

UNIT 33
How the Romans did Multiplication

Roman numerals are still used today but are not easy to calculate with and require a series of letters to represent numbers. The letters and the numbers they represent are shown in Figure 2.1.

The Romans used numbers mainly for counting so there was little need for a number symbol larger than 1000. Roman numbers were written by simply adding all the necessary symbols. For example, 2020 is written MMXX and 592 is written DLXXXXII. Later to avoid writing four symbols in a row, the idea of subtraction was employed by reversing the symbols. The number 592 could then be written as DXCII because the number 90 is written XC.

Addition of Roman numbers was manageable, but multiplication was more complex. It is not clear whether the method that was devised was completely understood by the Romans, but it is quite fascinating none the less. Suppose you want to multiply 36 by 105. You place the numbers in two columns and repeat the following steps:

Roman	I	V	X	L	C	D	M
Arabic	1	5	10	50	100	500	1000

Figure 2.1 Roman numbers

Take half of one number and double the other number. If half of a number is odd *round down* to a whole number. Stop when the halving process reaches the number 1. The process is shown below where the numbers on the right are continually halved and the numbers on the left are continually doubled.

36	105
72	52
144	26
288	13
576	6
1152	3
2304	1

Now cross out every number in the doubling column that is next to an even number in the halving column. Then sum up the remaining numbers in the doubling column and you have the product:

36	105
~~72~~	52
~~144~~	26
288	13
~~576~~	6
1152	3
2304	1
3780	

The product 36×105 is then 3780. If we reverse the columns the result is the same:

~~105~~	36
~~210~~	18
420	26
~~840~~	4
~~1680~~	2
3360	1
3780	

Here is what this looks like in Roman numerals:

XXXVI	CV
~~LXXII~~	LII
~~CXLIV~~	XXVI
CCLXXXVIII	XIII
~~DLXXVII~~	VI
MCDII	III
MMCCCIV	I
MMMDCCLXXX	

The Romans would add the numbers on the left by simply combining the symbols and substituting a symbol for a larger number when there are enough repeats of a symbol for a smaller number. For example, to double the first number above which is 36 you obtain:

XXXVI + XXXVI = XXXXXXVVII which then becomes LXXII.

We suspect you are now wondering: "How does this work?" First consider 2 numbers that are even and are both powers of two. Observe what happens when you apply the algorithm:

$$32 \quad 16$$

$$64 \quad 8$$

$$128 \quad 4$$

$$256 \quad 2$$

$$512 \quad 1$$

Note the following: The product of each pair of numbers in the two columns stays the same $32 \times 16 = 64 \times 8 = 128 \times 4 = 256 \times 2 = 512 \times 1 = 512$, which is the correct product. Now, since each number on the right is even, except the last one, all the numbers are crossed out, and the correct product is simply the last number in the doubles-column. The important idea here is that when we take half of an even number, the product of the two numbers does not change. The product does change, however when we take half of an odd number, which we show in the next example.

Consider again the first example above, and observe what happens to the product of each pair of numbers when you take half of an odd number and round down:

$$
\begin{array}{rclcl}
36 & \times & 105 & = & 3780 \\
\cancel{72} & \times & 52 & = & 3744 \quad -36 \\
\cancel{144} & \times & 26 & = & 3744 \\
288 & \times & 13 & = & 3744 \\
576 & \times & 6 & = & 3456 \quad -288 \\
1152 & \times & 3 & = & 3456 \\
\underline{2304} & \times & 1 & = & 3456 \\
\mathbf{3780}
\end{array}
$$

The first time you take half of an odd number and round down, the product decreases by half of the even number on the left as follows: Half of 105 is 52.5, which we round down to 52. Now the product of 72 times 52.5 is still the correct product. However, when we round down to 52, the product of 72 times 52 is now 3744, which is 36 less than the correct product of 3780. However, the number above 72 is 36 and it is next to an odd number. Therefore, it does not get crossed out and will be added at the end. By adding it at the end the reduction in the product is amended and preserves the correct result. Similarly, when we take half of the next odd number, 13, the product of the two numbers decreases by half of 576 which is 288. But 288 is next to an odd number and will also be added at the end, so the final result remains correct.

To summarize: *When a doubled even number that is next to an odd number is not crossed out, and added at the end to obtain the product, every decrease to the correct product is added back. Hence, the final result remains correct!*

Whoever discovered this unique Roman algorithm clearly was conversant in arithmetic and overcame a hurdle for the Romans enabling them to readily multiply their numbers. We hope you find this intriguing.

UNIT 34

How Complex Calculations were Done Before Electronics

The miracle of electronic calculation has significantly changed our world, mostly for the better, but has decreased our desire to do complex calculations mentally and by hand. However, for more than three thousand years, until only 60 years ago, people managed to do complex calculations using mathematical formulae and their mind. One's mental capacity, though it rivals any computer today, may eventually be surpassed by artificial intelligence which may be a mixed blessing.

One of the earliest advanced civilizations, the Babylonians, ca. 1500 BCE, had a very advanced number system, in some ways more advanced than ours. In the unit on Ancient Number Systems (page 1), it is shown that the Babylonian number system was based on 60, and not on 10 which is the base of our number system. However, it proved to be a very useful base, and their skillful measurement of time in hours, minutes and seconds still survives today. They developed superior calculating skills through the construction of tables, two of which provided the squares and cubes of numbers. One way they used the squares of numbers was to simplify multiplication by use of the formula:

$$a \times b = \frac{(a+b)^2 - (a-b)^2}{4}$$

For example, to multiply 42 by 27 this formula works as follows using base 10. First you find the sum and the difference of the two numbers, which are 69 and 15 respectively. Then using a table of

Figure 2.2   Replica of an ancient Abacus

squares, you obtain $69^2 = 4761$ and $15^2 = 225$. Now subtract these two numbers to produce 4536. Then divide 4536 by 4 to get the product, which is 1134. Division of large numbers was more difficult, but the Babylonians were able to do it by using an extensive table of reciprocals up to at least a million. To divide a by b, they would use their reciprocal table and multiply a by $\frac{1}{b}$.[1]

Around the same time as the Babylonian civilization flourished, the abacus came into existence, though it is not known exactly when. It was an expedient calculating tool that arose in China and found its way into parts of Asia and eventually into Europe. The abacus enabled the user to quickly do the 4 basic arithmetic operations, and also square roots and cube roots (see Figure 2.2).

It consists of several columns of movable beads which usually represent 1 or 5 digits. The beads are moved up or down to perform arithmetic operations. Though still in existence today in remote areas, it is gradually falling out of use.

Though the Babylonian civilization preceded the Greek and Roman civilizations, little advancement in calculating methods was developed until the beginning of the Renaissance. There were very

---

[1]See "The Saga of Mathematics", Lewinter & Widulski, Pearson 2001, Babylonian Arithmetic, p. 29.

talented mathematicians in Ancient Greece, but their interest was focused more on pure mathematical concepts rather than the simplification of complex calculations. For example, Euclid's *Elements* (ca. 300 BCE) was a brilliant treatise but contained mostly theoretical theorems and proofs. Archimedes (287–212 BCE) accomplished very intricate calculations, including an accurate approximation of the value of π by applying the *method of exhaustion*, a method which is the foundation for Integral Calculus. The method was developed a century earlier by Eudoxus of Cnidus (ca. 408–ca. 355 BCE). Archimedes also developed a method to calculate square roots accurately and invented a system for expressing large numbers. However, little of this found its way into everyday practical use with difficult calculations. The Romans, who succeeded the Greeks, were exceptional engineers and architects, but managed to do their calculations employing their less advanced number system. See the section on Roman Multiplication (page 109).

Major progress in mathematics lingered in Europe until the development of algebra by the Indian and Islamic civilizations during the 16th century. This culminated in the discovery of logarithms in the early 17th century by the Scottish mathematician John Napier (1550–1617). Subsequently, the British mathematician Henry Briggs (1561–1630) published tables of Napier's logarithms and promoted the acceptance of logarithms by scientists (see Figure 2.3).

Tables of logarithms enabled most learned people to do challenging calculations by reducing multiplication to addition and reducing division to subtraction. It also enabled exponential calculations and extracting roots of numbers by reducing exponentiation to multiplication and reducing root extraction to division. The slide rule, which is actually an analog computer, was developed around this time employing logarithmic scales. It is used primarily for multiplication and division but also can be used to do calculations involving exponents, roots, logarithms and trigonometric functions, but cannot be used for addition or subtraction (see Figure 2.4).

Slide rules are only accurate to 3 digits, while the accuracy of tables is limited by the number of digits used. Certain calculations involving many digits, such as large numbers associated with

min	Sinus	Logarithmi	Differentia	Logarithmi	Sinus	
0	0	Infinitum	Infinitum	0	10000000	60
1	2909	81425681	81425680	1	10000000	59
2	5818	74494213	74494211	2	9999998	58
3	8727	70439560	70439560	4	9999998	57
4	11636	67562746	67562739	7	9999993	56
5	14544	65331315	65331304	11	9999989	55
6	17453	63508099	63508083	16	9999986	54
7	20362	61966595	61966573	22	9999980	53
8	23271	60631284	60631256	28	9999974	52
9	26180	59453453	59453418	35	9999967	51
10	29088	58399857	58399814	43	9999959	50
11	31997	57446759	57446707	52	9999950	49
12	34906	56576646	56576584	62	9999940	48
13	37815	55776222	55776149	73	9999928	47
14	40724	55035148	55035064	84	9999917	46
15	43632	54345225	54345129	96	9999905	45
16	46541	53699843	53699734	109	9999892	44
17	49450	53093600	53093577	123	9999878	43
18	52359	52522019	52521881	138	9999863	42
19	55268	51981356	51981202	154	9999847	41
20	58177	51468431	51468361	170	9999831	40
21	61086	50980537	50980450	187	9999813	39
22	63995	50515342	50515137	205	9999795	38
23	66904	50070827	50070603	224	9999776	37
24	69813	49645239	49644995	244	9999756	36
25	72721	49237030	49236765	265	9999736	35
26	75630	48844826	48844539	287	9999714	34
27	78539	48467431	48467122	309	9999692	33
28	81448	48103763	48103431	332	9999668	32
29	84357	47752859	47752503	356	9999644	31
30	87265	47413852	47413471	381	9999619	30

89

Figure 2.3   Page from Napier's Logarithmic Tables 1614

astronomy, still required tedious effort to accomplish by hand. However, from the renaissance up to the late 19th century, before electricity was harnessed, many mechanical calculating machines were

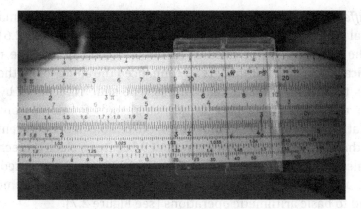

Figure 2.4   Slide rule

Figure 2.5   Early mechanical calculator invented by Pascal 1645

invented that were able to do basic arithmetic operations faster than by hand (see Figure 2.5).

One of the more advanced mechanical calculators was developed in 1820 by the British mathematician Charles Babbage (1791–1871).

His *difference engine* could automatically compute and print mathematical tables and tabulate polynomial functions (see Figure 2.6).

The development of calculus in the 17th century was the most significant achievement that greatly enhanced the ability of mathematicians to do very difficult calculations. The slide rule, aided by the power of calculus, provided the principal tools for doing complex calculations from the 17th century until the middle of the 20th century. With the discovery of electricity in the 19th century, electro-mechanical calculating machines evolved in the 1920's and increased the speed and accuracy of computations to 10 digits but were mostly limited to basic arithmetic operations (see Figure 2.7).

However, in the early 20th century mechanical calculating machines and devices continued to be used such as those shown in Figures 2.8 and 2.9.

Light bulbs were invented in 1879 and vacuum tubes, one of the first electronic components, were invented in 1906. In 1946, the first

Figure 2.6   Charles Babbage's difference engine

Figure 2.7 Electro-mechanical calculating machine

Figure 2.8 Mechanical calculator

electronic computer, the ENIAC which used 18,000 Vacuum Tubes and weighed 30 tons, was built by the United States Army. This was followed by the discovery of the transistor in 1947, and a whole new era dawned. Rapid development of mainframe computers ensued, followed by mini-computers and then personal computers which have become ubiquitous. Today, personal computers have been reduced to

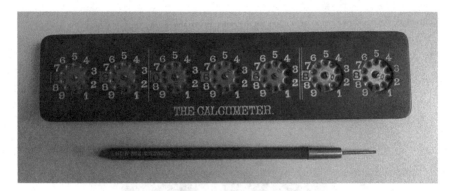

Figure 2.9

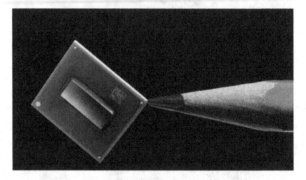

Figure 2.10   A microprocessor today

the size of a wristwatch and a tiny microprocessor has been developed by Intel, which is comparable to planning a city so miniscule it could fit into a single bacterium (see Figure 2.10).

We all marvel at the power and speed of electronic computers and calculators, and they are slowly closing the door on three-thousand years of human mental calculation, which, with much concern, could be heading toward extinction.

UNIT 35

American and European Subtraction Algorithms

The American method of subtraction may not always be the easiest algorithm; however, it is an algorithm that is very easily understood.

Just for review, let us take a look at the American algorithm to do the subtraction 627–135, and then compare it to a European algorithm.

Beginning with the units digit we have $7 - 5 = 2$, which provides the units digit of our difference. In the tens digit we would have to subtract $2 - 3$, which is not possible in the positive numbers, therefore, we borrow 10 tens (or 1 hundred) from the hundreds place so that we now have $12 - 3 = 9$, which belongs in the tens place of our difference. When we borrowed 10 hundreds from the hundreds place, we were left with 5 hundreds. We can complete our subtraction by subtracting in the hundreds position to get $5 - 1 = 4$ which is then placed in the hundreds position of our difference, yielding our complete difference of 492.

$$
\begin{array}{r}
5\,1\,2 \\
\not{6}\,\not{2}\,7 \\
-1\,3\,5 \\
\hline
4\,9\,2
\end{array}
$$

Many European countries use a subtraction algorithm often referred to as the *Austrian method of subtraction*. This is done rather differently from the American system, shown above. We will demonstrate the Austrian method of subtraction using the same two numbers as we used above, namely 627–135.

Once again, we begin with the units digits by asking, what number plus 5 will yield 7? This is, of course, 2, and we place the 2 in the units digit of the difference.

$$
\begin{array}{r}
6\,{}^{1}2\,7 \\
{}^{1}1\ 3\,5 \\
\hline
4\ 9\,2
\end{array}
$$

As before, we ask what number plus 3 will give us a 2? Since there is no number in the positive realm, we ask ourselves which number plus 3 will yield a 12? If we add a 10 to the 2 to get 12 and then ask ourselves the same question, we would find that $9 + 3 = 12$. Since we just added 10 tens to the subtrahend, we must also add the same amount to the minuend; this time in the form of 100s in order to keep this subtraction proper. Thus, we add 1 to the hundreds place, making $1 + 1 = 2$, so that we now ask what number plus 2 will yield a 6? So that we can then place a 4 in the hundreds place of our difference. Thus, we once again have a difference of 492.

We present this to show how various algorithms can be used to do subtraction. However, in today's world an electronic calculator would probably be the process of choice.

UNIT 36

Extracting a Square Root

Why would anyone want to find the square root of a number without using a calculator? Surely, no one would do such a thing. However, it is possible that there are some folks who are curious to know how it was done before calculators and computers were available and what is the procedure for extracting a square root. This could provide a better understanding of the process and also allow some independence from the calculator. We will use a method that was generally not taught in the schools but gives a good insight into the meaning of a square root. This method was first published in 1690 by the English mathematician Joseph Raphson (1668–1715) in his book, *Analysis aequationum universalis*, who attributed it to Newton, and therefore, the algorithm bears both names, the *Newton–Raphson method*.

It is perhaps best to see this method of square root extraction used in a specific example: Suppose we wish to find $\sqrt{27}$. (Obviously, the calculator could be used here and render us an immediate result of 1.196152423...) Using the Newton–Raphson method, we guess at what this value might be. Certainly, it is between $\sqrt{25}$ and $\sqrt{36}$, or between 5 and 6, but closer to 5.

Suppose we guess at 5.2. If this were the correct square root, then if we were to divide 27 by 5.2, we would get 5.2. But this is not the case here, since $\frac{27}{5.2} = 5.192$. Since $27 \approx 5.2 \times 5.192$, one of the factors

(5.2 in this case) must be larger than $\sqrt{27}$ and the other factor (5.192 in this case) must be less than $\sqrt{27}$. Hence, $\sqrt{27}$ is sandwiched between the two numbers 5.2 and 5.192, that is, $5.192 < \sqrt{27} < 5.2$. Therefore, it is plausible to infer that the average, that is, $\frac{5.2+5.192}{2} = 5.196$ is a better approximation for $\sqrt{27}$ than either 5.2 or 5.192.

This division process continues: $\frac{27}{5.192} = 5.196$, so each time we obtain an additional decimal place, which leads to a closer approximation. That is, $\frac{5.192+5.196}{2} = 5.194$, then $\frac{27}{5.1940} = 5.19831$. Taking this process another step further we take the average as before, to get: $\frac{5.194+5.19831}{2} = 5.196155$. If we want to get an additional decimal place accuracy, we would continue the process as follows: $\frac{27}{5.196155} = 5.1961498$ and then $\frac{5.196155+5.1961498}{2} = 5.1961524$. As we progress, we are getting ever closer to the value of $\sqrt{27}$, which we can check on the calculator. This continuous process provides an algorithm for finding the square root of a number, which is not a perfect square. Cumbersome as the method may be, it gives some insight into what a square root represents.

UNIT 37

Divisibility by Prime Numbers

It is often useful to be able to look at a number and easily determine if it is divisible by certain numbers. We can directly determine when a number is divisible by 2 or 5 as follows: If the last digit of the number in question is an even number, then the number is divisible by 2, and if a number ends in a 0 or 5, then the number is divisible by 5. The calculator certainly allows us to detect which numbers divide into a given number but there are clever ways to determine without a calculator whether a given number is divisible by prime numbers other than 2 or 5. To better understand and appreciate mathematics, divisibility techniques provide an interesting "window" into the nature of numbers and their properties. For this reason (among others), the topic of divisibility still finds a place on the mathematics-learning spectrum.

Divisibility by 3 (or 9)

There are moments in everyday situations where the factors of a number can be useful to know, especially if it can be done mentally. For example, a restaurant bill of $71.22 needs to be split into 3 equal parts. Before actually doing the division, the thought about whether or not it is possible to split the bill equally may come into question. Wouldn't it be nice if there were some mental arithmetic shortcuts for determining this? There is a method to determine if a number is divisible by 3 and (as an extra bonus) divisible by 9. The technique is:

> *If the sum of the digits of a number is divisible by 3 (or 9),*
> *then the original number is divisible by 3 (or 9).*[2]

Perhaps an example would best firm up an understanding of this technique. Consider the number 296,357 which we shall test for divisibility by 3 (or 9). The sum of the digits is $2 + 9 + 6 + 3 + 5 + 7 = 32$, which is not divisible by 3 or 9. Therefore, the original number 296,357 is not divisible by 3 nor 9. Another example: Is the number 457,875 divisible by 3 or 9? The sum of the digits is $4 + 5 + 7 + 8 + 7 +5 = 36,$[3] which is divisible by 9 (and then by 3 as well), so the number 457,875 is divisible by 3 and by 9.

Certainly, there are times when a number is divisible by 3 and not by 9 as is the case with the number 27,987. The sum of the digits is $2 + 7 + 9 + 8 + 7 = 33$, which is divisible by 3 but not by 9; therefore,

---

[2]*For the interested reader, here is a brief discussion about why this rule works as it does.

Consider the number ab, cde, whose value can be expressed as

$N = 10^4 a + 10^3 b + 10^2 c + 10d + e = (9 + 1)4a + (9 + 1)3b + (9 + 1)2c + (9 + 1)d + e = [9M + (1)4]a + [9M + (1)3]b + [9M + (1)2]c + [9 + (1)]d + e = 9M [a + b + c + d] + a + b + c + d + e$, which implies that divisibility by 9 of N depends on the divisibility of: $a + b + c + d + e$, the sum of the digits.

Note: 9M refers to a multiple of 9.

[3] If by some remote chance it is not immediately clear to you if the sum of the digits is divisible by 3 or 9, then take the sum of the digits of this number and continue the process until you can visually make a determination of its divisibility by 3 or 9.

the number 27,987 is divisible by 3 and not by 9. We can go back to the original question about the divisibility of the restaurant bill of $71.22. Can it be divided into three equal parts? Because $7 + 1 + 2 + 2 = 12$, and 12 is divisible by 3, then $71.22 is divisible by 3.

It has always been quite challenging to establish techniques for divisibility by other prime numbers. This is especially true with the technique for divisibility by 7, as its method can also establish a procedure for the remaining prime numbers beyond 7. Although some techniques to determine divisibility might be a bit cumbersome, they are fun, and believe it or not, can occasionally prove to be useful. Let us consider this technique for divisibility by 7, and then as we inspect it see how this can be generalized for other prime numbers.

Divisibility by 7

Delete the last digit from the given number, and then subtract twice this deleted digit from the remaining number. If the result is divisible by 7, the original number is divisible by 7. This process may be repeated if the result is too large for simple inspection of divisibility of 7.

Suppose we want to test the number 876,547 for divisibility by 7. Begin with 876,547 and delete its units digit, 7, and subtract its double, 14, from the remaining number: $87,654 - 14 = 87,640$. Since we cannot yet visually inspect the resulting number for divisibility by 7, we continue the process. Continue with the resulting number 87,640 and delete its units digit, 0, and subtract its double, still 0, from the remaining number; we get: $8,764 - 0 = 8,764$. Since this did not change the resulting number, 8,764, as we seek to check for divisibility by 7, we continue the process. Continue with the result, 8,764, delete its units digit, 4, and subtract its double, 8 from the remaining number. We then obtain: $876 - 8 = 868$. Since we still cannot visually inspect the resulting number, 868, for divisibility by 7 we continue the process. Continuing with the last number, 868, we delete its units digit, 8, and subtract its double, 16, from the remaining number. We now have: $86 - 16 = 70$, which is clearly divisible by 7. Therefore, the number 876,547 is divisible by 7.

Terminal digit	Number subtracted from original	Terminal digit	Number subtracted from original
1	20 + 1 = 21 = 3×7	5	100 + 5 = 105 = 15×7
2	40 + 2 = 42 = 6×7	6	120 + 6 = 126 = 18×7
3	60 + 3 = 63 = 9×7	7	140 + 7 = 147 = 21×7
4	80 + 4 = 84 = 12×7	8	160 + 8 = 168 = 24×7
		9	180 + 9 = 189 = 27×7

Figure 2.11

Now let's reveal the beauty of this algorithm! To justify the technique of determining divisibility by 7, consider the various possible terminal digits (that you are "dropping") and the corresponding subtraction that is being done by dropping the last digit. In Figure 2.11, you will see how dropping the terminal digit and doubling gives us in each case a multiple of 7. That is, we have taken "bundles of 7" away from the original number. Therefore, if the remaining number is divisible by 7, then so is the original number, because you have separated the original number into two parts, each of which is divisible by 7, and therefore, the entire number must be divisible by 7.

For example, supposing we would like to determine if the number 3628 is divisible by 7. Delete the 8 and double it to get 16 and when we subtract 16 from the remaining number, we have actually subtracted 168 from the original number, which is a multiple of 7 (168 = 24 × 7). This leaves us with the number 346. Repeating this process, we delete the 6 and subtract 12 from 34 to get 22, which is not divisible by 7. Therefore, we can conclude that the original number 3628 is not divisible by 7.

Divisibility by 11

At the oddest times, the issue can come up to determine if a number is divisible by 11. If you have a calculator at hand, the problem is easily solved. But that is not always the case. Besides, there is such a clever technique for testing for divisibility by 11 that it is worth knowing just for its charm. The technique is quite simple:

*If the difference of the sums of the alternate digits is divisible
by 11, then the original number is also divisible by 11.*

It may sound a bit complicated, but it really isn't. Let us take this
technique one step at a time. The sum of the alternate digits means
that you begin at one end of the number taking the first, third, fifth,
etc. digits and add them. Then add the remaining (even placed) digits.
Subtract the two sums and inspect for divisibility by 11.

It is best demonstrated through an example. We shall test 768,614
for divisibility by 11. Sums of the alternate digits are: $7 + 8 + 1 = 16$,
and $6 + 6 + 4 = 16$. The difference of these two sums is $16 - 16 = 0$,
which is divisible by 11.[4] Therefore, we can conclude that 768,614 is
divisible by 11.

Divisibility by 13

*This is similar to the technique for testing divisibility
by 7, except that the 7 is replaced by 13 and instead of subtracting twice the deleted digit, we subtract nine times the
deleted digit each time.*

Let's check for divisibility by 13 for the number 5,616. Begin
with 5,616 and delete its units digit, 6, and subtract $9 \times 6 = 54$, from
the remaining number: $561 - 54 = 507$. Since we still cannot visually
inspect the resulting number for divisibility by 13, we continue the
process with the resulting number 507 and delete its units digit and
subtract $9 \times 7 = 63$ from the remaining number: $50 - 63 = -13$,
which is divisible by 13, and therefore the original number is divisible by 13.

To determine the "multiplier" 9, we sought the smallest multiple
of 13 that ends in a 1. That was 91, where the tens digit is 9 *times* the
units digit. Once again consider the various possible terminal digits
and the corresponding subtractions shown in Figure 2.12.

---

[4]Remember $\frac{0}{11} = 0$

Terminal digit	Number subtracted from original	Terminal digit	Number subtracted from original
1	$90 + 1 = 91 = 7 \times 13$	5	$450 + 5 = 455 = 35 \times 13$
2	$180 + 2 = 182 = 14 \times 13$	6	$540 + 6 = 546 = 42 \times 13$
3	$270 + 3 = 273 = 21 \times 13$	7	$630 + 7 = 637 = 49 \times 13$
4	$360 + 4 = 364 = 28 \times 13$	8	$720 + 8 = 728 = 56 \times 13$
		9	$810 + 9 = 819 = 63 \times 13$

Figure 2.12

In each case a multiple of 13 is being subtracted one or more times from the original number. Hence, if the remaining number is divisible by 13, then the original number is divisible by 13.

Divisibility by 17

Delete the units digit and subtract five times the deleted digit each time from the remaining number until you reach a number small enough to determine its divisibility by 17.

We justify the technique for divisibility by 17 as we did the techniques for 7 and 13. Each step of the procedure subtracts a "bunch of 17s" from the original number until we reduce the number to a manageable size and can make a visual inspection of divisibility by 17.

The patterns developed in the preceding three divisibility techniques (for 7, 13, and 17) should lead you to develop similar techniques for testing divisibility by larger primes. Figure 2.13 presents the "multipliers" of the deleted digits for various primes.

You may want to extend your knowledge of divisibility techniques to include composite (i.e. non-prime) numbers. The technique for divisibility by composite numbers is:

A given number is divisible by a composite number if it is divisible by each of its relatively prime[5] factors.

---

[5] Two numbers are relatively prime if they have no common factors other than 1.

To test divisibility by	7	11	13	17	19	23	29	31	37	41	43	47
Multiplier	2	1	9	5	17	16	26	3	11	4	30	14

Figure 2.13

To be divisible by	6	10	12	15	18	21	24	26	28
The number must be divisible by	2, 3	2, 5	3, 4	3, 5	2, 9	3, 7	3, 8	2, 13	4, 7

Figure 2.14

Figure 2.14 offers illustrations of this technique.

By providing a rather comprehensive list of techniques for testing divisibility, we hope this gives you an interesting insight into elementary number theory.

UNIT 38

Successive Percentages

In our everyday experiences, we come across times when a store offers a discount on top of a previously discounted item. Typically, we tend to add the two discounts and compute the amount discounted based on that percentage sum. It turns out that this is incorrect. We will consider a process that is not only quite useful but also, to some extent, counterintuitive. We begin by considering the following problem:

> Wanting to buy a coat, Barbara is faced with a dilemma. Two competing stores next to each other carry the same brand coat with the same list price, but with two different discount offers. Store A offers a 10% discount year-round on all its goods, but on this particular day offers an additional 20% on top of their already discounted price. Store B simply offers a discount of 30% on that day in order to stay competitive. Is there a difference in the price? If so, which is the better offer?

At first glance, you may assume that there is no difference in price, since $10 + 20 = 30$, yielding the same discount in both cases. Yet, with a little more thought you may realize that this is not correct, since in store A only 10% is calculated on the original list price, with the 20% calculated on the lower price, while at store B, the entire 30% is calculated on the list price. Now, the question to be answered is, what percentage difference is there between the discount in store A and in store B? One expected procedure might be to assume the cost of the coat to be $100. Calculating the 10% discount yielding a $90 price, and an additional 20% of the $90 price (or $18) will bring the price down to $72. In store B, the 30% discount on $100 would bring the price down to $70, giving a discount difference of $2, which in this case is 2%. This procedure, although correct, and not too difficult, is a bit cumbersome and does not always allow full insight into the situation.

An interesting and quite unusual procedure requires a fresh look into this problem and provides some entertainment as well. Here is the mechanical method for obtaining a single percentage discount (or increase) equivalent to two (or more) successive discounts (or increases).

(1) *Change each of the percentages involved into decimal form*:

$$0.20 \text{ and } 0.10$$

(2) *Subtract each of these decimals from 1.00*:

$$0.80 \text{ and } 0.90 \text{ (for an increase, add to 1.00)}$$

(3) *Multiply these differences*:

$$(0.80)(0.90) = 0.72$$

(4) *Subtract this number (i.e. 0.72) from 1.00*:

$1.00 - 0.72 = 0.28$. *This represents the combined discount.*

(If the result of step 3 is greater than 1.00, subtract 1.00 from it to obtain the percent of increase.)

When we convert 0.28 back to percent form, we obtain 28%, the equivalent of successive discounts of 20% and 10%.

This combined percentage of 28% differs from 30% by 2%.

Following the same procedure, you can also combine more than 2 successive discounts. In addition, successive increases, combined or not combined, with a discount, can also be done using this procedure. By adding the decimal equivalent of the increase to 1.00, while subtracting the decimal equivalent of the discount from 1.00 and then continuing the procedure in the same way. If the end result turns out to be greater than 1.00, then this end result reflects an overall increase rather than the discount as was found in the above problem. This procedure not only streamlines a typically cumbersome situation, but also provides some insight into the overall picture. For example, the question "Is it advantageous to the buyer in the above problem to receive a 20% discount and then a 10% discount, or the reverse: 10% discount and then a 20% discount?" The answer to this question is not immediately intuitively obvious. Yet, since the procedure just presented shows that the calculation is merely multiplication, which is a commutative operation, there is no difference between the two methods.

So here you have a delightful algorithm for combining successive discounts or increases or combinations of these. Not only is it useful, but it also gives you some newly found power in dealing with percentages, when using a calculator is not available.

UNIT 39

Casting Out Nines

We will use the number 9 to demonstrate a hidden mathematical trick that can be used to check ordinary arithmetic. Essentially, it will allow us to determine if our arithmetic computation results could be correct.

Before we discuss this arithmetic-checking procedure, we will consider how the remainder of a division by 9 compares to removing groups of 9s from the digit sum of the number. Let's begin by finding the remainder, when 8,768 is divided by 9. The quotient is 974 with a remainder of 2. We now show that this remainder can also be obtained by "casting out nines" from the digit sum of the number 8,768. This means that we first find the sum of the digits and if the sum is more than a single digit, we repeat the procedure of taking digit sums until we reach a single digit. In the case of our given number, 8,768, the digit sum is $8 + 7 + 6 + 8 = 29$. Since we have not reached a single digit, we repeat the process. Now the digit sum is now: $2 + 9 = 11$, so again to reach a single digit we get the digit sum of 11 which is $1 + 1 = 2$, and which turns out to be the same as the remainder when we divided 8,768 by 9. We call this procedure "casting-out-nines" because we are removing groups of 9 from the number, which can be done either by constant divisions or by taking the sum of the digits as we have done here.

Let's see how this technique can assist us in checking to see if a multiplication example could be correct. Consider the product $734 \times 879 = 645,186$. We can check this by division (Is $645,186 \div 734 = 879$?), but that would be somewhat lengthy if we do not have a calculator at hand. However, we can check if this is correct by using our "casting out nines" procedure as follows: Take the digit sum of each factor and the product. Continue this until a single digit number is reached.

For 734:	$7 + 3 + 4 = 14$;	then $1 + 4 = \mathbf{5}$
For 879:	$8 + 7 + 9 = 24$;	then $2 + 4 = \mathbf{6}$
For 645,186:	$6 + 4 + 5 + 1 + 8 + 6 = 30$,	then $3 + 0 = \mathbf{3}$.

Since $\mathbf{5 \times 6 = 30}$, which yields 3 (the digit sum of 30 is $3 + 0 = 3$), which is the same as the digit sum for the product (645,186), the answer could be correct. We say "could be correct" because if the digits were interchanged the result would be the same. Therefore, it is not a perfect check of the arithmetic procedure.

For practice we will do another casting-out-nines "check" for the following multiplication:

$$56,589 \times 983,678 = 55,665,354,342$$

For 56,589: $5 + 6 + 5 + 8 + 9 = 33;$ $\Rightarrow 3+3 = 6$

For 983,678: $9 + 8 + 3 + 6 + 7 + 8 = 41;$ $\Rightarrow 4+1 = 5$

For 55,665,354,342: $5 + 5 + 6 + 6 + 5 + 3 + 5 + 4 + 3 + 4 + 2 = 48;$ $\Rightarrow 4+8 = 12$

$\Rightarrow 1+2 = 3$

To check for possibly having the correct product: $6 \times 5 = 30$ or $3 + 0 = 3$, which matches the 3 resulting from the digit sum of the product (55,665,354,342).

The same procedure can be used to check the likelihood of a correct sum, difference, or quotient, simply by following the same procedure as above, that is, taking the groups of 9 out of each of the numbers being added, subtracted, or divided and combining these digit sums with the same arithmetic procedure and comparing that result to the digit sum of the sum, difference, or quotient, respectively. We will consider an example of casting-out-nines for each of the other three arithmetic processes. For addition, consider the following problem:

$6845 \Rightarrow 6+8+4+5 = 23 \Rightarrow 2+3 = \boxed{5}$

$2618 \Rightarrow 2+6+1+8 = 17 \Rightarrow 1+7 = \boxed{8}$ $5+8+9 = 22 \Rightarrow 2+2 = \boxed{4}$

$\underline{+3654} \Rightarrow 3+6+5+4 = 18 \Rightarrow 1+8 = \boxed{9}$

$13117 \Rightarrow 1+3+1+1+7 = 13 \Rightarrow 1+3 = \boxed{4}$

In each case, we took the sum of the digits and then continued taking sums until a single digit was reached. The sum of the three-digit sums equals 22 whose digit sum is 4. This is equal to the digit sum of the total, indicating that the answer could have been correct. Once again, by removing groups of 9 from each of the numbers and considering only the remainder, which is the same taking the sum of the digits continuously until a single digit is arrived at. This is the remainder when dividing the original number by 9. Working with the remainders is one form of checking to see if the sum could be correct.

Bear in mind that a different digit arrangement of the sum would still get the same result and the answer would not be correct.

Applying the casting out nines process to a subtraction example, we will consider the following:

$$9657 \Rightarrow 9+6+5+7=27 \Rightarrow 2+7= \boxed{9}$$
$$-3284 \Rightarrow 3+2+8+4=17 \Rightarrow 1+7= \boxed{8} \quad 9-8 = \boxed{1}$$
$$6373 \Rightarrow 6+3+7+3=19 \Rightarrow 1+9= 10 \Rightarrow 1+0= \boxed{1}$$

Here we notice that the difference of the digit sums of the minuend and the subtrahend (the number being subtracted) is 1, which matches the sum of the digits of the difference. This once again indicates that the subtraction could be correct.

Lastly, we will apply the casting-out-nines procedure to a division problem, namely $563{,}753 \div 44 = 12{,}812$ remainder 25. Once again, when we apply the casting-out-nines procedure, we place the result in a box. Follow below as we use the procedure to see if our calculation could be correct.

$$12812 \Rightarrow 1+2+8+1+2=14 \Rightarrow 1+4=\boxed{5} \quad \text{Remainder} =25 \Rightarrow 2+5=\boxed{7}$$
$$44\overline{)563753} \Rightarrow 5+6+3+7+5+3=29 \Rightarrow 2+9=11 \Rightarrow 1+1=\boxed{2}$$
$$\Downarrow$$
$$4+4=\boxed{8}$$

To check the division, we multiply the quotient (12812) by the divisor (44), which we will now do with the digit sums (5 and 8) of these numbers after we had cast out the nines: $5 \times 8 = 40$, and then we will add the digit sum (7) of the remainder:

$$40+7=47 \Rightarrow 4+7=11 \Rightarrow 1+1=\boxed{2}$$

Because this corresponds to the digit sum (2) of the dividend (563753), the division could be correct. It should be stressed that this procedure of casting-out-nines is merely a test to see if the arithmetic calculation could be correct. It does not say that it *is* correct. However, it does give us further insight into the arithmetic processes and the power of the number 9. We once again, attribute this procedure to Leonardo of Pisa (Fibonacci) who in 1202 included it in his book *Liber abaci.*

UNIT 40

Division by Zero

Every mathematician knows that division by zero is *forbidden*. As a matter of fact, on the list of "commandments" in mathematics this must certainly be at the top. But why is division by zero not permissible? In mathematics, there is a pride that the order and beauty in which everything in the realm of mathematics falls neatly into place and furthermore is justifiable. When something arises that could spoil that order, we simply *define* it to suit our needs. This is precisely what happens with division by zero. One gets a much greater insight into the nature of mathematics by explaining why these "rules" are set forth. So, let's give this "commandment" some meaning.

Consider the quotient $\frac{n}{0}$, where n is not equal to zero. Without acknowledging the division-by-zero commandment, let us speculate (i.e. guess) what the quotient might be. Let us say it is p, so that $\frac{n}{0} = p$. In that case, we could check by multiplying $0 \times p$ to see if it equals n as would have to be the case for the division to be correct. We know that $p \times 0 \neq n$, since $0 \times p = 0$. So, there is no number, p that can take on the quotient to this division. Because we have the rule that any number times zero is equal to zero, it follows that a number divided by zero cannot be defined.

Now we have the reason why division by zero is not permitted. There are mathematics teachers who drive this point home by telling their students that the 11th commandment is: "thou shalt not divide by zero." Here is a fun algebraic example of how one can prove that $1 = 2$, which results from a division by 0:

$$\text{Let } a = b$$
$$\text{Then } a^2 = ab \qquad \text{[multiplying both sides by } a]$$
$$a^2 - b^2 = ab - b^2 \qquad \text{[subtracting } b^2 \text{ from both sides]}$$
$$(a - b)(a + b) = b(a - b) \qquad \text{[factoring both sides]}$$
$$a + b = b \qquad \text{[dividing by both sides by } (a - b]$$
$$2b = b \qquad \text{[replace } a \text{ with } b, \text{ since } a = b]$$
$$2 = 1 \qquad \text{[divide both sides by } b]$$

In the step, where we divided both sides of the equation by $(a - b)$, we actually divided by zero, because $a = b$, so $a - b = 0$. That ultimately led us to an absurd result, namely, that $1 = 2$, thereby, leaving us with no option other than to prohibit division by zero.

UNIT 41

Mistaken Assumptions

It can be very tempting to let lots of consistent examples lead you to a generalization. Often a generalization may be correct, but it doesn't have to be. The famous mathematician Carl Friedrich Gauss (1777–1855) was known to have used his brilliance at calculating and mentally processing number relationships to form some of his theories. Then he proved his conjectures and his contributions to the field of mathematics have become legendary. You must be cautious not to draw conclusions just because lots of examples fit a pattern. It is necessary to *prove* mathematically that your assumption is true for any example.

The French mathematician Alphonse de Polignac (1817–1890) stated that "every odd number greater than 1 can be expressed as the sum of a power of 2 and a prime number." If we inspect the first few cases, we find that this appears to be a true statement. However, as you will see from the list in Figure 2.15, it holds true for the odd

Odd number	Sum of a power of 2 and a prime number
3	$= 2^0 + 2$
5	$= 2^1 + 3$
7	$= 2^2 + 3$
9	$= 2^2 + 5$
11	$= 2^3 + 3$
13	$= 2^3 + 5$
15	$= 2^3 + 7$
17	$= 2^2 + 13$
19	$= 2^4 + 3$
...	...
51	$= 2^5 + 19$
...	...
125	$= 2^6 + 61$
127	$= ?$
129	$= 2^5 + 97$
131	$= 2^7 + 3$

Figure 2.15

numbers from 3 through 125 and then is *not* true for 127; after which it continues to hold true again for a while.

The next numbers that fail de Polignac's conjecture are 149, 251, 331, 337, 373, and 509, while another counterexample is 877. Consider the number 149 where this conjecture does not work. This can be verified as follows. If there is a power of 2 less than 2^8 + a prime number that equals 149 then we need to verify that. We begin with $2^8 = 256 > 149$, we simply need to verify that $149 - 2^k$ is not prime for every value of $k = 1$ to 7. We can check this as follows:

$149 - 2^0 = 149 - 1 = 148$ (divisible by 2)
$149 - 2^1 = 149 - 2 = 147$ (divisible by 3)
$149 - 2^2 = 149 - 4 = 145$ (divisible by 5)
$149 - 2^3 = 149 - 8 = 141$ (divisible by 3)
$149 - 2^4 = 149 - 16 = 133$ (divisible by 7)
$149 - 2^5 = 149 - 32 = 117$ (divisible by 3)
$149 - 2^6 = 149 - 64 = 85$ (divisible by 5)
$149 - 2^7 = 149 - 128 = 21$ (divisible by 3)

You can do the same for other counterexamples to see that they also fail to have any kind of decomposition into 2^k + prime, since we can check all possible values of 2^k and see that the difference is never prime. Thus, this cannot be generalized. Caution should be taken from jumping to conclusions, especially when no proof has been developed. This is a good example of drawing premature conclusions.

By the way, in 1849, Alphonse de Polignac proposed another conjecture that to date has not been proved or disproved. It is as follows: *There are infinitely many differences of length n between two consecutive prime numbers.*

For example, suppose we let $n = 2$. There are a few smaller consecutive prime number pairs whose difference is 2, such as (3, 5), (11, 13), (17, 19), etc. Again, we still have not established if this conjecture is true or false. Remember, we can only accept something as true if it can be proved so!

On the other hand, there are some simple mistakes we make and because they lead to absurdities, we tend to dismiss them without much thought. We know that if equals are multiplied by equals, then the results are equal. For example, if we know that $x = y$, then we also can conclude that $3x = 3y$. Yet, when we have 2 pounds = 32 ounces, and $\frac{1}{2}$ pound = 8 ounces, then by multiplying equals, does $2 \times \frac{1}{2}$ pound = 32×8 ounces? Which implies that 1 pound = 256 ounces? Of course not. Where did we go wrong?

Similarly, we know that $\frac{1}{4}$ dollar = 25 cents. Then by taking the square root of both sides one gets $\sqrt{\frac{1}{4} \text{dollar}} = \sqrt{25}$ cents, which is then $\frac{1}{2}$ dollar = 5 cents? Again, absurd! Where did we go wrong this time? When we multiplied the numbers or took their square roots, we didn't consider that the units were different, which led us to an incorrect solution — albeit an awkward one! To make this explanation a bit simpler, suppose we begin with 2 feet = 24 inches, and $\frac{1}{2}$ foot = 6 inches, then by multiplying the units with their corresponding measures, we get:

$2 \times \left(\frac{1}{2}\right) \times (\text{foot})^2 = 24 \times (6) \times (\text{inches})^2$ or 1 square foot = 144 square inches, which is correct!

Thus, caution must be used in generalizing things that have not been proved and making some false assumptions.

UNIT 42

The Ulam–Collatz Loop

There are certain experiences in mathematics that tend to be inexplicable. Some feel that when something is truly surprising and "neat", it is beautiful. From that standpoint, we will show a seemingly "magical" property in mathematics. This is one that has baffled mathematicians for many years and still no one knows why it happens. Does this amazing scheme that we are about to expose *really* work for *all* numbers? This loop-generating scheme was first discovered in 1932 by Lothar Collatz (1910–1990), a German mathematician, who then published it in 1937. Credit is also given to the American mathematician, Stanislaus Marcin Ulam (1909–1984), who worked on the Manhattan Project during World War II.

To demonstrate this amazing relationship, we begin by providing the following two rules that will guide us through the process:

> *If the number is odd, then multiply by 3 and add 1.*
>
> *If the number is even, then divide by 2.*

One can begin with any *arbitrary* number and regardless of the number selected, after continued repetition of the process, the resulting number will always be 1.

Let's try it for the *arbitrary* number **7**:

7 is odd, therefore, multiply by 3 and add 1 to get: $7 \times 3 + 1 = \mathbf{22}$.

22 is even, so we simply divide by 2 to get **11**.

11 is odd, so we multiply by 3 and add 1 to get **34**.

34 is even, so we divide by 2 to get **17**.

17 is odd, so we multiply by 3 and add 1 to get **52**.

52 is even, so we divide by 2 to get **26**.

26 is even, so we divide by 2 to get **13**.

13 is odd, so we multiply by 3 and add 1 to get **40**.

40 is even, so we divide by 2 to get **20**.

20 is even, so we divide by 2 to get **10**.
10 is even, so we divide by 2 to get **5**.
5 is odd, so we multiply by 3 and add 1 to get **16**.
16 is even, so we divide by 2 to get **8**.
8 is even, so we divide by 2 to get **4**.
4 is even, so we divide by 2 to get **2**.
2 is also even, so we again divide by 2 to get **1**.

If we were to continue, we would find ourselves in a loop, as you can see that continuing with the number 1, we see that since 1 is odd, we multiply by 3 and add 1 to get **4**, and then it follows the same path as the last few steps above. Thus, we have entered into a loop.

Therefore, we get the number sequence 7, 22, 11, 34, 17, 52, 26, 13, 40, 20, 10, 5, 16, 8, **4, 2, 1, 4, 2, 1**, ...

The following schematic provided in Figure 2.16, will show the path we have just taken: The vertical paths show tripling the number and adding one, the horizontal paths show merely halving the number.

Regardless which number we begin with (here we started with **7**), we will eventually get to 1, and continuing that process results in a loop. This is truly remarkable! Try it for some other numbers to convince yourself that it really does work. Had we started with **9** as our arbitrarily selected number, it would have required **19** steps to

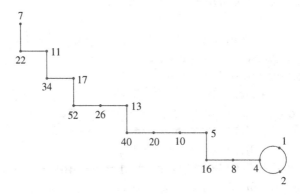

Figure 2.16

reach 1. Starting with **41** will require 109 steps to reach 1. One of the amazing features of this scheme is that no matter which number we begin with, we always end up with the following loop of numbers of length 3: [4, 2, 1].

A proof that this holds for all numbers has not yet been found. The Canadian mathematician, H. S. M. Coxeter (1907–2003), offered a prize of $50 to anyone who could come up with such a proof, and $100 for anyone who could find a number for which this doesn't work. Later, the Hungarian mathematician, Paul Erdös (1913–1996), raised the prize money to $500. Still, with all these and many further incentives, no one has yet found a proof. This seemingly "true" algorithm then must remain a conjecture until it is proved true for all cases.

Most recently (with aid of computers), this "$3n + 1$ Problem," as it is also commonly known, has been shown to hold true for the numbers up to $18 \times 2^{58} \approx 5.188146770 \times 10^{18}$ (June 1, 2008)[6] — that means, for more than 5 Quintillion numbers it has been shown to be true.

UNIT 43

A Cyclic Number Loop

In the previous section, we considered a number loop which is a circular arrangement that is quite unique and entertaining. It also gives further insights into numbers. Consider the following cyclic number loop. We begin by taking any integer from 1 to 6 and multiplying it by 999,999 and then dividing it by 7. You will get a number made up of the digits 1, 4, 2, 8, 5, 7. Not only that, but they will be in this order, yet starting from a different digit each time. This is the phenomenon of a *cyclic number*; which is several *n* digits that when

---

[6] See http://www.ieeta.pt/~tos/3x+1.html. Also, see Tomás Oliveira e Silva, *Maximum excursion and stopping time record-holders for the 3x+1 problem: Computational results*, Mathematics of Computation, Vol. 68, No. 225, pp. 371–384, 1999.

multiplied by each of the numbers 1, 2, 3, 4,...., n, produces a number that uses the same digits as the original number, but in a different order each time.

These numbers are sometimes also called *Phoenix numbers* — after the bird that according to an ancient Egyptian legend rises youthfully from its ashes whenever it is burned and disappears (see Figure 2.17).

If we multiply the number 142,857 by 7, we get 999,999. Another peculiarity occurs when we multiply the number 142,857 by 8; we get 1,142,856. With some imagination, we could take the first digit and add it to the last digit and then we would be back to where we started with the number 142,857. Taking this a step further, when we multiply 142,857 by 9, we get 1,285,713. Using the same awkward technique by taking the first digit and adding it to the last digit, we get the product of the number 142,857 when multiplied by 2.

There is still more we can show with this unusual number. Not only is the sum of the digits of each of the products 27, but if we take the sum of the digits vertically, we will also get a sum for each place value to be 27, as shown in Figure 2.18.

We now come to a very peculiar aspect of this curious number. Suppose we multiply this number, 142,857, by another large number, say, the number 32,789,563,521. We get the product: 4,684,**218, 675,**919,497. We will break up this product in groups of six beginning at the right side of the number and then add the numbers.

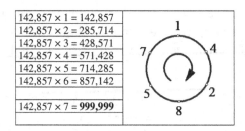

Figure 2.17

```
1  4  2  8  5  7 = 27
2  8  5  7  1  4 = 27
4  2  8  5  7  1 = 27
5  7  1  4  2  8 = 27
7  1  4  2  8  5 = 27
8  5  7  1  4  2 = 27
───────────────────
27 27 27 27 27 27
```

Figure 2.18

919,497
218,675
  4,684
─────────
1,142,856

We are almost finished with this demonstration. All we now need to do is to take these "six-groupings" one step further: We will take the first 1 and add it to the remaining number to get the following:

142,856
      1
─────────
142,857

By now you are probably not surprised to see that the number 142,857 appears again. Just to demonstrate this as not being "rigged," we will do this with another product of 142,857 and the large number 89,651,273,582,410,598 to get: **12,807**,311,990,**162,430**,798,486. Then breaking this number up into groups of six, beginning on the right side, we get the following:

798,486
162,430
311,990
 12,807
─────────
1,285,713

Once again, we will take the first 1 and add it to the number again, and we notice the almost-anticipated result.

$$
\begin{array}{r}
285,713 \\
1 \\
\hline
285,714 = 142,857 \times 2
\end{array}
$$

A motivated reader may want to try other such products of 142,857 to verify this phenomenon. This is merely one example of a cyclic situation which occurs often in mathematics.

UNIT 44

Curious Number Properties

Mathematicians would like to think that all experiences in mathematics can be explained. This is not always the case, as we can see with several amazing curiosities that exist in our number system. Let's explore just a few of these to whet your appetite.

Curious numbers with unit-digit 9

There are patterns that we can discover such as the following:

$$19 = 1 \times 9 + (1 + 9)$$
$$29 = 2 \times 9 + (2 + 9)$$
$$39 = 3 \times 9 + (3 + 9)$$
$$49 = 4 \times 9 + (4 + 9)$$
$$59 = 5 \times 9 + (5 + 9)$$
$$69 = 6 \times 9 + (6 + 9)$$
$$79 = 7 \times 9 + (7 + 9)$$
$$89 = 8 \times 9 + (8 + 9)$$
$$99 = 9 \times 9 + (9 + 9)$$

The Curious Number 11

First note that the number 11 is the only palindromic prime number that has an even number of digits. Here are some other curiosities with the number 11 to amuse you further about.

The square of 11 can be expressed as the sum of 5 consecutive powers of 3 as follows:

$11^2 = 121 = 3^0 + 3^1 + 3^2 + 3^3 + 3^4$. Furthermore, the cube of 11 can be expressed as the sum of the squares of 3 consecutive odd numbers: $11^3 = 1331 = 19^2 + 21^2 + 23^2$. We can also express the number 11 as the sum of a square and a prime in three different ways. (And, by the way, it is the smallest number that has this property.)

$$2^2 + 7 = 11$$

$$3^2 + 2 = 11$$

Now here is one that is really unusual. If we reverse the digits of any number, which is divisible by 11, the resulting number will also be divisible by 11. To demonstrate this, consider the number 135,916 $= 11 \times 12,356$. When you reverse the digits to get 619,531, it just happens to be $11 \times 56,321$, which clearly is also a multiple of 11. You may wish to try this with other numbers.

The Curious Number 110

When we tag a zero onto the number 11, we get the number 110. At first glance, there does not seem to be anything particularly unusual about that number. It belongs to a set of numbers known as *sphenic* numbers, which are numbers that are the product of 3 distinct prime numbers, in this case $110 = 2 \times 5 \times 11$. It also belongs to a set of numbers known as *pronic* numbers, which are numbers that are the product of 2 consecutive numbers, where in this case $110 = 10 \times 11$. However, the number 110 also has a more unique feature in that it can be represented as the sum of squares in precisely 3 different ways:

$$110 = 1^2 + 3^2 + 10^2 = 1 + 9 + 100$$

$$110 = 5^2 + 6^2 + 7^2 = 25 + 36 + 49$$

$$110 = 2^2 + 5^2 + 9^2 = 4 + 25 + 81$$

We have a similar situation for a 4-digit number, such as 8,208 which is equal to the sum of the fourth powers of its digits: $8,208 = 8^4 + 2^4 + 0^4 + 8^4$.

The Sum of Squares Surprise

If we take this a step further and consider any 4-digit number, the sum of squares of the digits will lead to a very interesting result. Suppose we begin with the randomly selected 4-digit number 1,527. We find that the sum of the squares of the digits is $1^2 + 5^2 + 2^2 + 7^2 = 1 + 25 + 4 + 49 = 79$. We now continue this process by taking the sum of the squares of the digits of the number $79 = 7^2 + 9^2 = 130$. Once again, repeating this process we get the following: $1^2 + 3^2 + 0^2 = 1 + 9 + 0 = 10$. The continued process then yields: $1^2 + 0^2 = 1$. What you will notice is that regardless of which number you select at the beginning, this process of taking the sum of the squares of the digits of each resulting number will either end up with the number 1, as was the case in our illustration above, or you will end up with the number 4, in which case the subsequently-generated numbers after 4 will be the repeating sequence: 4, 16, 37, 58, 89, 145, 42, 20, **4, 16, 37, 58, 89, 145, 42, 20**, 4, 16, 37, 58, 89, 145, 42, 20.

Squares That Can Be Partitioned into Squares

There are numbers whose square is comprised of digits that can be partitioned into two squares as follows:

$$7^2 = \quad 4\overline{9} = \underline{2}^2\,\overline{3}^2$$
$$13^2 = \quad 16\overline{9} = \underline{4}^2\,\overline{3}^2$$
$$19^2 = \quad 36\overline{1} = \underline{6}^2\,\overline{1}^2$$
$$35^2 = \quad 1\,\overline{225} = \underline{1}^2\,\overline{15}^2$$
$$38^2 = \quad 144\,\overline{4} = \underline{12}^2\,\overline{2}^2$$
$$57^2 = \quad 324\,\overline{9} = \underline{18}^2\,\overline{3}^2$$
$$223^2 = \underline{49}\,\overline{729} = \underline{7}^2\,\overline{27}^2$$

Splitting Numbers

Moving along to other number curiosities, consider the number 3025. Suppose we split the number as 30 and 25, and then add these two numbers to get $30 + 25 = 55$. Then, if we square this result, we get back to the original number: $55^2 = 3025$. The question that immediately arises is "Are there other numbers for which this little procedure also works?"

Well, here are a few more such numbers:

$$9801 \rightarrow 98 + 01 = 99 \qquad \text{and} \quad 99^2 = 9,801$$
$$2025 \rightarrow 20 + 25 = 45 \qquad \text{and} \quad 45^2 = 2,025$$
$$088209 \rightarrow 088 + 209 = 297 \qquad \text{and} \quad 297^2 = 88,209$$
$$494209 \rightarrow 494 + 209 = 703 \qquad \text{and} \quad 703^2 = 494,209$$
$$998001 \rightarrow 998 + 001 = 999 \qquad \text{and} \quad 999^2 = 998,001$$
$$99980001 \rightarrow 9998 + 0001 = 9999 \quad \text{and} \quad 9999^2 = 99,980,001$$

While we are in the mode of "splitting" numbers, let's partition numbers to get some other fascinating relationships:

$$1{,}233 \rightarrow 12|33 = 12^2 + 33^2$$

$$8{,}833 \rightarrow 88|33 = 88^2 + 33^2$$

$$990{,}100 \rightarrow 990|100 = 990^2 + 100^2$$

$$94{,}122{,}353 \rightarrow 9412|2353 = 9412^2 + 2353^2$$

$$7{,}416{,}043{,}776 \rightarrow 74160|43776 = 74160^2 + 43776^2$$

$$116{,}788{,}321{,}168 \rightarrow 116788|321168 = 116788^2 + 321168^2$$

$$221{,}859 \rightarrow 22|18|59 = 22^3 + 18^3 + 59^3$$

$$166{,}500{,}333 \rightarrow 166|500|333 = 166^3 + 500^3 + 333^3$$

You might like to verify some of the following that can be split, where the sum of the squares of the parts will equal the original number:

$$10{,}100 \rightarrow 10|100 = 10^2 + 100^2,$$

$$588{,}2353 = 588|2353 = 5882 + 2353^2.$$

Here are a few more such numbers: 99010|09901, 17650|38125, 25840|43776, 999900|010000, 123288|328768, and there are still more!

An analogous situation — this time using differences instead of sums — also produces some interesting relationships:

$$48 \rightarrow 4|8 = 8^2 - 4^2$$

$$3{,}468 \rightarrow 34|68 = 68^2 - 34^2$$

$$416{,}768 \rightarrow 416|768 = 768^2 - 416^2$$

$$33{,}346{,}668 \rightarrow 3334|6668 = 6668^2 - 3334^2$$

Combining Powers

A further charming occurrence in mathematics is seen with the following illustrations, which highlight the notion that there are probably limitless nuggets we can cull from our number system.

$16^3 + 50^3 + 33^3 = 4{,}096 + 125{,}000 + 35{,}937 = \mathbf{165033} = 165{,}033.$

$166^3 + 500^3 + 333^3 = 4{,}574{,}296 + 125{,}000{,}000 + 36{,}926{,}037 = \mathbf{166,}$ $\mathbf{500, 333} = 166{,}500{,}333.$

$588^2 + 2353^2 = 345{,}744 + 5{,}536{,}609 = \mathbf{5{,}88\ 2{,}353} = 5{,}882{,}353.$

Powers of Digits

Let's just look at the numbers 135 and 175. On the surface you would think they have nothing in common. Well, they are two of the numbers that share a very special property. Look what happens when each of their digits is raised to a power one greater than the previous digit.

Notice that each of these numbers equals the sum of their digits raised to its consecutive exponents.

$$135 = 1^1 + 3^2 + 5^3$$
$$175 = 1^1 + 7^2 + 5^3$$
$$518 = 5^1 + 1^2 + 8^3$$
$$598 = 5^1 + 9^2 + 8^3$$

It is natural to ask if there are 4-digit numbers that also have this amazing property. Here are some that satisfy this relationship.

$$1306 = 1^1 + 3^2 + 0^3 + 6^4$$
$$1676 = 1^1 + 6^2 + 7^3 + 6^4$$
$$2427 = 2^1 + 4^2 + 2^3 + 7^4$$

Now if you thought these were unusual numbers, you will probably be quite enchanted with the next number property. It is really amazing. Notice the relationship between the exponents and the numbers.[7]

$$3435 = 3^3 + 4^4 + 3^3 + 5^5$$
$$438579088 = 4^4 + 3^3 + 8^8 + 5^5 + 7^7 + 9^9 + 0^0 + 8^8 + 8^8$$

Who said mathematics doesn't have its "beauties" to show off?

Five Unique Numbers

Can you imagine that a number is equal to the sum of the cubes of its digits? This is true for only five numbers. Below are these five most unusual numbers.

$$1 \rightarrow 1^3 = 1$$
$$153 \rightarrow 1^3 + 5^3 + 3^3 = 1 + 125 + 27 = 153$$
$$370 \rightarrow 3^3 + 7^3 + 0^3 = 27 + 343 + 0 = 370$$
$$371 \rightarrow 3^3 + 7^3 + 1^3 = 27 + 343 + 1 = 371$$
$$407 \rightarrow 4^3 + 0^3 + 7^3 = 64 + 0 + 343 = 407$$

Take a moment to appreciate these spectacular results and take note that these are the *only* such numbers for which this is true.

---

[7] In the second example the expression 0^0 is defined by mathematicians to be indeterminate, yet for simplicity sake (and to make our example work) we shall give it a value of 0.

More Number Curiosities

As an "extension" of the Armstrong Numbers (see page 79), we will present numbers where the sum of the digits of the original number — each taken to the power one more or less than the number of digits in the original number — is equal to the original number. This differs from the Armstrong numbers, where the power to which each of the digits was taken was equal to the number of digits in the original number.

To begin, consider the following 4-digit numbers where each of the digits is take to the 5th power, and the sum is equal to the original number.

$$4150 = 4^5 + 1^5 + 5^5 + 0^5$$

$$4151 = 4^5 + 1^5 + 5^5 + 1^5$$

The following are some other examples of non-Armstrong numbers:

$$194979 = 1^5 + 9^5 + 4^5 + 9^5 + 7^5 + 9^5$$

$$14459929 = 1^7 + 4^7 + 4^7 + 5^7 + 9^7 + 9^7 + 2^7 + 9^7$$

Notice how in these cases, quite the opposite of the previous numbers, the powers reflect the original number, and the base stays the same:

$$4,624 = 4^4 + 4^6 + 4^2 + 4^4$$

$$1,033 = 8^1 + 8^0 + 8^3 + 8^3$$

$$595968 = 4^5 + 4^9 + 4^5 + 4^9 + 4^6 + 4^8$$

$$3909511 = 5^3 + 5^9 + 5^0 + 5^9 + 5^5 + 5^1 + 5^1$$

$$13177388 = 7^1 + 7^3 + 7^1 + 7^7 + 7^7 + 7^3 + 7^8 + 7^8$$

$$52135640 = 19^5 + 19^2 + 19^1 + 19^3 + 19^5 + 19^6 + 19^4 + 19^0$$

Then there are those numbers that differ from the previous examples and lend themselves to an even more astonishing pattern: here the powers and the base match the original digits:

$$3435 = 3^3 + 4^4 + 3^3 + 5^5$$
$$438579088 = 4^4 + 3^3 + 8^8 + 5^5 + 7^7 + 9^9 + 0^0 + 8^8 + 8^8$$

We can also find a reversal of this pattern, namely, where the bases go in the proper order and the exponents progress in the reverse order:

$$48625 = 4^5 + 8^2 + 6^6 + 2^8 + 5^4$$
$$397612 = 3^2 + 9^1 + 7^6 + 6^7 + 1^9 + 2^3$$

Searching through our number system reveals that there are also some entertaining patterns where consecutive exponents are used:

$$43 = 4^2 + 3^3$$
$$63 = 6^2 + 3^3$$
$$89 = 8^1 + 9^2$$
$$1676 = 1^5 + 6^4 + 7^3 + 6^2$$

Yet, when we formalize this a bit and use the consecutive natural numbers as exponents, we have the following amazing relationships:

$$135 = 1^1 + 3^2 + 5^3$$
$$175 = 1^1 + 7^2 + 5^3$$
$$518 = 5^1 + 1^2 + 8^3$$
$$598 = 5^1 + 9^2 + 8^3$$
$$1306 = 1^1 + 3^2 + 0^3 + 6^4$$
$$1676 = 1^1 + 6^2 + 7^3 + 6^4$$
$$2427 = 2^1 + 4^2 + 2^3 + 7^4$$
$$2646798 = 2^1 + 6^2 + 4^3 + 6^4 + 7^5 + 9^6 + 8^7$$

You may have noticed that 1,676 appeared in both lists of consecutive exponents.

The Devil Number

The remarkable number, 666, which is often referred to as the devil number — and considered "unlucky" — seems to almost have a boundless array of numerical "coincidences" embedded within it, one being that the sum of the numbers on a roulette wheel is 666.

Here are a few delectable number relationships, which lead to 666:

$$666 = 1^6 - 2^6 + 3^6$$

$$666 = (6 + 6 + 6) + (6^3 + 6^3 + 6^3)$$

$$666 = (6^4 - 6^4 + 6^4) - (6^3 + 6^3 + 6^3) + (6 + 6 + 6)$$

$$666 = 5^3 + 6^3 + 7^3 - (6 + 6 + 6)$$

$$666 = 2^1 \times 3^2 + 2^3 \times 3^4$$

We can even generate 666 by representing each of its three digits in terms of 1, 2, and 3:

$$6 = 1 + 2 + 3$$

$$6 = 1 \times 2 \times 3$$

$$6 = \sqrt{1^3 + 2^3 + 3^3}$$

Therefore, $666 = (100)(1 + 2 + 3) + (10)(1 \cdot 2 \cdot 3) + \left(\sqrt{1^3 + 2^3 + 3^3} \right)$

Number coincidences

Lastly here are some numbers that share a rather curious relationship as follows:

$$1^1 + 6^1 + 8^1 - 15 - 2^1 + 4^1 + 9^1$$
$$1^2 + 6^2 + 8^2 = 101 = 2^2 + 4^2 + 9^2$$

$$1^1 + 5^1 + 8^1 + 12^1 = 26 = 2^1 + 3^1 + 10^1 + 11^1$$
$$1^2 + 5^2 + 8^2 + 12^2 = 234 = 2^2 + 3^2 + 10^2 + 11^2$$
$$1^3 + 5^3 + 8^3 + 12^3 = 2{,}366 = 2^3 + 3^3 + 10^3 + 11^3$$

$$1^1 + 5^1 + 8^1 + 12^1 + 18^1 + 19^1 = 63 = 2^1 + 3^1 + 9^1 + 13^1 + 16^1 + 20^1$$
$$1^2 + 5^2 + 8^2 + 12^2 + 18^2 + 19^2 = 919 = 2^2 + 3^2 + 9^2 + 13^2 + 16^2 + 20^2$$
$$1^3 + 5^3 + 8^3 + 12^3 + 18^3 + 19^3 = 15{,}057 = 2^3 + 3^3 + 9^3 + 13^3 + 16^3 + 20^3$$
$$1^4 + 5^4 + 8^4 + 12^4 + 18^4 + 19^4 = 260{,}755 = 2^4 + 3^4 + 9^4 + 13^4 + 16^4 + 20^4$$

This is just the beginning of considering number curiosities. There are many more to be found in books such as *Numbers: Their Tales, Types, and Treasures* by A. S. Posamentier and B. Thaller (*Prometheus Books*, 2015).

UNIT 45

Magic Squares

Mathematical puzzles serve to stimulate mathematical reasoning, and therefore form an important part of recreational mathematics. They are often analogous to crossword puzzles, but the entries are numbers instead of words. A popular number-related puzzle is Sudoku (Japanese for "single number"), a logic-based number-placement puzzle that originated in Japan and gained world-wide popularity in 2005. A similar but somewhat less-popular puzzle, yet with a closer relation to arithmetic operations than Sudoku, is known under the name Kenken or KenDoku, which is logical number-placement puzzle.

One of the first puzzles in the history of mathematics has been the magic square, and today it is as fascinating as it was ages ago. The task is to find a square arrangement of numbers, such that the sum of the numbers in each row and each column is equal to the sum of the numbers in each of the two diagonals. The first known example of a magic square is the Lo Shu square, with numbers arranged as shown in Figure 2.18. It was known to Chinese mathematicians as early as 650 BCE, and became important in Feng Shui, the art of placing objects to achieve harmony with the surrounding environment.

A legend from that time tells us that once there was a huge flood on the Lo River in China, and the people tried to placate the river's god. But each time they offered a sacrifice, a turtle emerged from the river, and walked around the offering until a child noticed a strange pattern of dots on the turtle's shell. After studying these markings, the people realized that the correct number of sacrifices to make would be 15. And after they did so, the river god was to have been satisfied and as a result the flood receded. The number 15 is the sum of numbers in each row, column, and diagonal of the Lo Shu magic square (Figure 2.19), which was pictured on the turtle.

$$
\begin{array}{cccccc}
1 & 4 & 2 & 8 & 5 & 7 = 27 \\
2 & 8 & 5 & 7 & 1 & 4 = 27 \\
4 & 2 & 8 & 5 & 7 & 1 = 27 \\
5 & 7 & 1 & 4 & 2 & 8 = 27 \\
7 & 1 & 4 & 2 & 8 & 5 = 27 \\
8 & 5 & 7 & 1 & 4 & 2 = 27 \\
\hline
27 & 27 & 27 & 27 & 27 & 27 \\
\end{array}
$$

Figure 2.18   Lo Shu square and the magic turtle

4	9	2
3	5	7
8	1	6

Figure 2.19   Lo Shu square and the magic turtle

Magic squares appear throughout history and were particularly popular among Arabic mathematicians in Baghdad, who even designed 6 × 6 magic squares, and published them in an encyclopedia in 983 CE. In the 10th century, a famous magic square, called Chautisa Yantra, appeared in India. The 4 × 4 magic square shown in Figure 2.20, is found on the Parshvanath temple in Khajuraho, India. For this magic square the sum of each row, each column, and the diagonals is 34.

There is one magic square however that stands out from among the rest for its beauty and additional properties — not to mention its curious appearance. This particular magic square has many properties beyond those required for a typical magic square, and comes to us through art, and not through the usual mathematical channels. It is depicted in the background of the famous 1514 engraving entitled *Melencolia* by the renowned German artist Albrecht Dürer (1471–1528), who lived in Nürnberg, Germany (see Figure 2.21).

Remember, a magic square is a square matrix of numbers, where the sum of the numbers in each of its columns, rows, and diagonals is the same. As we begin to examine the magic square in Dürer's etching, we should take note that most of Dürer's works were signed by him with his initials, one over the other, and with the year in which the work was made included. Here we find it in the dark-shaded region near the lower right side of the picture (Figure 2.21 and in detail Figure 2.22). We notice that it was made in the year 1514.

7	12	1	14
2	13	8	11
16	3	10	5
9	6	15	4

Figure 2.20   Chautisa Yantra

Figure 2.21   Melencolia I, engraving, Albrecht Dürer (1514)

The observant reader may notice that the two center cells of the bottom row of the Dürer magic square depict the year as well. Let us examine this magic square more closely (see Figure 2.23).

First let's make sure that it is in fact a true magic square. When we find the sum of each of the rows, columns and diagonals, we get the sum of 34, which is all that is be required for this square matrix of numbers to be considered a "magic square". However, this "Dürer

Figure 2.22   Initials AD of Albrecht Dürer and the year 1514

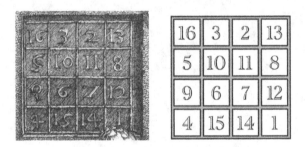

Figure 2.23   Dürer's magic square

Magic Square" has many more properties than other magic squares Let us now marvel over some of these extra properties.

- The four corner numbers have a sum of 34:
  $16 + 13 + 1 + 4 = 34$
- Each of the four corner 2 by 2 squares has a sum of 34
  $16 + 3 + 5 + 10 = 34$
  $2 + 13 + 11 + 8 = 34$
  $9 + 6 + 4 + 15 = 34$
  $7 + 12 + 14 + 1 = 34$

- The center 2 by 2 square has a sum of 34
  $10 + 11 + 6 + 7 = 34$
- The sum of the numbers in the diagonal cells equals the sum of the numbers in the cells not in the diagonal:
  $16 + 10 + 7 + 1 + 4 + 6 + 11 + 13 = 3 + 2 + 8 + 12 + 14 + 15 + 9 + 5 = 68$
- The sum of the squares of the numbers in both diagonal cells is
  $16^2 + 10^2 + 7^2 + 1^2 + 4^2 + 6^2 + 11^2 + 13^2 = 748$
  This number 748 is equal to

  - the sum of the squares of the numbers *not* in the diagonal cells.
    $3^2 + 2^2 + 8^2 + 12^2 + 14^2 + 15^2 + 9^2 + 5^2 = 748$
  - the sum of the squares of the numbers in the first and third rows.
    $16^2 + 3^2 + 2^2 + 13^2 + 9^2 + 6^2 + 7^2 + 12^2 = 748$
  - the sum of the squares of the numbers in the second and fourth rows.
    $5^2 + 10^2 + 11^2 + 8^2 + 4^2 + 15^2 + 14^2 + 1^2 = 748$
  - the sum of the squares of the numbers in the first and third columns.
    $16^2 + 5^2 + 9^2 + 4^2 + 2^2 + 11^2 + 7^2 + 14^2 = 748$
  - the sum of the squares of the numbers in the second and fourth columns.
    $3^2 + 10^2 + 6^2 + 15^2 + 13^2 + 8^2 + 12^2 + 1^2 = 748$

- The sum of the cubes of the numbers in the diagonal cells equals the sum of the cubes of the numbers not in the diagonal cells:
  $16^3 + 10^3 + 7^3 + 1^3 + 4^3 + 6^3 + 11^3 + 13^3 =$
  $= 3^3 + 2^3 + 8^3 + 12^3 + 14^3 + 15^3 + 9^3 + 5^3 = 9{,}248$
- Notice the following beautiful symmetries:
  $2 + 8 + 9 + 15 = 3 + 5 + 12 + 14 = 34$
  $2^2 + 8^2 + 9^2 + 15^2 = 3^2 + 5^2 + 12^2 + 14^2 = 374$
  $2^3 + 8^3 + 9^3 + 15^3 = 3^3 + 5^3 + 12^3 + 14^3 = 4624$
- Adding the first row to the second, and the third row to the fourth, produces a pleasing symmetry:

$16 + 5 = \mathbf{21}$	$3 + 10 = \mathbf{13}$	$2 + 11 = \mathbf{13}$	$13 + 8 = \mathbf{21}$
$9 + 4 = \mathbf{13}$	$6 + 15 = \mathbf{21}$	$7 + 14 = \mathbf{21}$	$12 + 1 = \mathbf{13}$

▪ Adding the first column to the second, and the third column to the fourth, produces a pleasing symmetry:

$16 + 3 = 19$	$2 + 13 = 15$
$5 + 10 = 15$	$11 + 8 = 19$
$9 + 6 = 15$	$7 + 12 = 19$
$4 + 15 = 19$	$14 + 1 = 15$

A motivated reader may wish to search for other patterns in this beautiful magic square. Remember, this is not a typical magic square, where all that would be required is that all the rows, columns and diagonals have the same sum. This Dürer magic square has many more properties.

The questions that would logically be asked is how can one construct a magic square? How did Dürer come up with this special magic square? According to their classifications, we can distinguish three types of magic squares:

(a) magic squares of odd order (n is an odd number),

(b) magic squares of doubly-even order (n is a multiple of 4),

(c) magic squares of singly-even order (n is a multiple of 2, but not 4).

The Dürer magic square is a doubly-even magic square.

There are many sources that provide methods for creating these various categories of magic squares. One possible source is: *Numbers: Their Tales, Types and Treasures,* by A. S. Posamentier and B. Thaller. Amherst, NY: *Prometheus Books*, 2015, pp. 191–206.

Chapter 3

Aspects of Measurement

UNIT 46

The Origin of Measurement

The history of attempts to establish accurate measurements has gone through a myriad of manifestations. The earliest forms of measurement were efforts to assign numbers to physical quantities, and therefore constituted the beginning of mathematics. Ancient weight measurement was based on objects such as stones or beans, and length measurement parts of the human body, usually a royal one, such as the length of the foot or the spread of the hand. One of the earliest basic units devised by the Egyptians was the cubit, which was the length of the forearm from the elbow to the end of the fingers. The smallest unit was the digit which was the width of a finger and there were 28 digits in a cubit. Other units included the palm, the hand and the small cubit which was 24 digits. As a result, the earliest problems in mathematics were concerned with weights and measures. These methods for assigning units continued throughout the Greek and Roman civilizations. The basic unit of the Greeks was the breadth of a finger (about 19 mm), with 16 fingers in a foot and 24 fingers in a cubit. The Greeks also adopted the Egyptian units of weight and volume. The Romans used the foot as their basis and divided it into

12 inches, while longer units were based on marching. Five feet was equal to a double step and was named a pace.

A Roman mile contained 1,000 paces (2000 steps) and so the Roman mile was about 5,000 feet. The Romans marked off these miles with stones along the many roads they built in Europe — hence the name, "milestones." The name *statute mile* goes back to Queen Elizabeth I of England who redefined the mile from 5,000 feet to 8 furlongs[1] (5,280 feet) by a statute in 1593.

As the Roman Empire expanded, its measurement system became entrenched throughout Europe, with the exception of England, as it was occupied by many different peoples who brought other units into play such as the furlong and the fathom.[2] Measurement systems were not well standardized and there was little uniformity or accuracy throughout the world until the 18th century. Finally, in 1790, the French proposed a more scientific system of international measurement based on the decimal system. The basic unit was the meter defined as the length of a pendulum that took one second to complete a cycle, because the beat of a pendulum depends only on its length. However, the beat does differ at different latitudes, and while the French proposed 45°, which was in France, the British proposed London, and the United States proposed the 38th parallel, which was close to Thomas Jefferson's estate. The meter was redefined in 1793 as one ten-millionth of the distance from the equator to the North

---

[1] A furlong is a measure of distance within Imperial units and U.S. customary units. Although its definition has varied historically, in modern terms it equals 660 feet, and is therefore equal to 201.168 meters. There are eight furlongs in a mile. The name "furlong" derives from the Old English words furh (furrow) and lang (long). It originally referred to the length of the furrow in one acre of a ploughed open field (a medieval communal field which was divided into strips). The term is used today for distances horses run at a racetrack.

[2] A fathom is equal to 6 feet (1.8288 m) and is primarily used for measuring the depth of water. It is historically the most frequently employed maritime measure of depth in the English-speaking world. Originally the span of a man's outstretched arms, the size of a fathom has varied slightly depending on whether it was defined as a thousandth of a nautical mile (1 nm = 1.151 statute miles) or as a multiple of the imperial yard.

Pole along a great circle such as the Earth's circumference which is approximately 40,000 km. Eventually in September 1799, a platinum bar was constructed, and it became the official definition of the meter. Still many countries resisted adopting the meter until 1812 when Napoleon's conquests helped spread the use of the system and it became more widely used throughout Europe. It still took many years for most countries to uniformly adopt the French system.

The U.S. Customary system evolved from the old English System, which has its roots in the Ancient Roman, Carolingian and Saxon units of measure. The U.S. system was developed after the American Revolution taking its core from English units employed in the thirteen colonies. The U.S. units are defined in terms of metric units where 1 yard = 0.9144 meters, and 1 pound = 453.592 grams. By an Act of Congress in 1866, it became "lawful throughout the United States to employ the weights and measures of the metric system in all contracts, dealings, or court proceedings." However, there is no such law establishing the use of our mile system.

In 1960, the 11th General Conference on Weights and Measures established the International System of Units, universally abbreviated SI (from the French *Le Système International d'Unités*), as the official measuring system. As was expected, it took many years for the SI system to be universally adopted. At the same time, the physical meter bar in Paris was replaced by a more precise standard based on the wavelength of light produced by the Krypton-86 atom, and then in 1983 it was redefined as the distance light travels in a vacuum in $\frac{1}{299,792,458}$ seconds, which is its present definition.

The British Empire developed the Imperial System in 1826, and it was used until 1985 when England gradually replaced it with the SI system. The U.S. Customary system has remained for various reasons, mainly economic, but there were also some who believed that the metric system was atheistic. In 1975, President Ford signed the Metric Conversion Act into law which stated that the metric system is "the preferred system of weights and measures for United States trade and commerce" but allowed all activities to continue to use the U.S. Customary system, and it is still used more extensively than the metric system. Presently three countries, Myanmar, Liberia, and the United States, have not officially adopted the metric system, how-ever, it is employed in many industries in these countries including the medical, electrical, military, and automotive fields. This does pres-ent a dilemma for the average consumer as measurements on prod-ucts can be stated in either system, and one needs to be conversant in both systems to better understand and compare the measurement associated with an item. See the following unit *Metric System versus the U.S. Customary System.*

UNIT 47

The Metric System Versus the U.S. Customary System

If you grew up with the United States Customary measurement sys-tem and were very comfortable with it, the metric system might seem awkward at first and not easy to envision the quantities that the units measure. The metric system is a decimal system (based on 10) and is therefore, more logical mathematically than the U.S. system, which has no clear base. However, when one is comfortable using a system, it can serve one better than having to learn a new system, no matter how logical. A good example are the units used to measure time which, though they are over 4 thousand years old and based on 60, were not changed in the SI system. They had sufficed for most of the world for all those years, and it would have proven too cumbersome to base the units of time on 10 by, perhaps, break-ing the hour into 100 minutes and the day into 20 hours. When President Ford tried to switch the United States to the international or SI system, in 1975, some oil companies started selling gasoline by the Liter and consumers were so confused by the prices because they seemed so low (1 gallon = 0.264 liters), that they had to switch back to the U.S. system. Therefore largely for economic reasons, the United States continues to embrace an antiquated system of measurement.

You may wonder if there are any measurements in the U.S. system more logical than the equivalent measurements in the metric system. Well, one can argue that perhaps the Fahrenheit scale used to measure temperature is more comfortable to work with than the Celsius or centigrade scale. The Fahrenheit scale was developed by the Dutch physicist Daniel Gabriel Fahrenheit (1686–1736). By most accounts, he set 0°F as the freezing temperature of a solution of water and a salt, ammonium chloride, and 100°F as the average human body temperature, which was only approximately determined at that time. The Celsius scale was developed by the Swedish astronomer Anders Celsius (1701–1744) who designated 0°C as the freezing point of water and 100°C as the boiling point of water.

Now, consider the Fahrenheit range of daily temperatures which most people experience in the world. The coldest temperatures hover near 0°F and the hottest temperatures near 100°F. One could submit that this range of numbers from 0 to 100 is based on 10 and is comfortable to work with. The equivalent Celsius range of daily temperatures is between −18°C and 38°C and might not be considered as friendly a range of numbers as 0 to 100 since it includes negative values. Daniel Fahrenheit chose values based on the coldest and warmest temperatures most people experience throughout the world and set the useful range of his scale from 0 to 100, which, one can argue, makes more logical sense than the equivalent Celsius temperatures. The bottom line is, of course, whatever you have learned in your early years and with which you have become most accustomed, will probably be the most comfortable values to work with.

In the United States and other countries, one can be confronted with measurements in either system and it is useful to have a convenient way to compare units. Here are some approximate comparisons that can be helpful:

A meter is a little longer than a yard.

A Liter is a little larger than a quart.

An inch is approximately 2½ centimeters.

A kilogram is a little more than 2 pounds.

A convenient way to quickly change Celsius to Fahrenheit, is to double the temperature in Celsius and add 30. This is an approximate calculation that can be done quickly with an error of 4° or less. The actual formula to convert from Celsius to Fahrenheit is:

$$F = 1.8°C + 32$$

In another country, where kilometers are used instead of miles, a quick conversion can be done with specially designed calculators or by some "trick" method. That is where the Fibonacci numbers come in. Recall the Fibonacci numbers: 1, 1, 2, 3, 5, 8, 13, 21, 34, 55, 89,144,... where each number is the sum of the two preceding numbers.

Now to convert from miles to kilometers and vice versa, we need to see how one mile relates to the kilometer. The statute mile (our usual measure of distance in the United States today) is exactly 1609.344 meters long which is 1.609344 kilometers. On the other hand, one kilometer is .621371192 miles long. The nature of the last two numbers, which are reciprocals that differ by almost 1, might remind us of the golden ratio, which is approximately 1.618 and its reciprocal is approximately 0.618 (see page 223). Remember the golden ratio is the only number whose reciprocal differs from it by exactly 1. This would tell us that the Fibonacci numbers, the ratio of whose consecutive members approaches the golden ratio, might come into play here.

Let's see what length 5 miles would have in kilometers: 5 × 1.609344 = 8.04672 which is approximately 8. We could also check to see what the equivalent of 8 kilometers would be in miles: 8 × 0.621371192 = 4.970969536, which is approximately 5. This allows us to conclude that approximately 5 miles is equal to approximately 8 kilometers. Here we have 2 of our Fibonacci numbers.

The relationship between miles and kilometers is very close to the golden ratio, as is the ratio of a Fibonacci number to the one before it, so using this relationship, we would be able to approximately *convert 8 kilometers to miles*, replace 8 by the previous Fibonacci number 5, and 8 kilometers is about 5 miles as shown above. Similarly, 5 kilometers is about 3 miles and 2 kilometers is about 1 mile. The higher Fibonacci numbers will give us a more accurate estimate, since the ratio of these larger consecutive Fibonacci numbers gets closer to the golden ratio.

Good luck with your metric immersion!

Now suppose you want to convert 20 kilometers to miles. We have selected 20, since it is *not* a Fibonacci number. We can express 20 as a *sum of Fibonacci numbers*[3] and convert each number separately and

---

[3] Representing Numbers as Sums of Fibonacci Numbers: It is not a trivial matter for us to conclude that every natural number can be expressed as the sum of other Fibonacci numbers without repeating any one of them in the sum. Let's take the first few Fibonacci numbers to demonstrate this property:

n	The Sum of Fibonacci Numbers Equal to n
1	1
2	2
3	3
4	1+3
5	5
6	1+5
7	2+5
8	8
9	1+8
10	2+8
11	3+8
12	1+3+8
13	13
14	1+13
15	2+13
16	3+13

You should begin to see patterns and also note that we used the fewest number of Fibonacci numbers in each sum in the figure above. For example, we could also have represented 13 as the sum of 2 + 3 + 8, or as 5 + 8. Try to express larger natural numbers as the sum of Fibonacci numbers. Each time ask yourself if you have used the fewest number of Fibonacci numbers in the sum. It will be fun to see the patterns that develop. By the way, Edouard Zeckendorf (1901–1983) proved, each natural number is a (unique) sum of nonconsecutive Fibonacci numbers.

then add them. Thus, 20 kilometers = 13 kilometers + 5 kilometers + 2 kilometers, which, by replacing 13 by 8, and 5 by 3, and 2 by 1, is approximately equal to 12 miles. To use this process to achieve the *reverse*, that is, to convert miles to kilometers, we write the number of miles as a sum of Fibonacci numbers and then replace each by the next *larger* Fibonacci number. Let's convert 20 miles to kilometers. 20 miles = 13 miles + 5 miles + 2 miles. Now, replacing each of the Fibonacci numbers with the next larger in the sequence, we get: 20 miles = 21 kilometers + 8 kilometers + 3 kilometers = 32 kilometers.

There is no need to use the Fibonacci representation of a number that uses the fewest Fibonacci numbers. You can use any combination of numbers whose sum is the number you are converting. For instance, 40 kilometers is 2 × 20 and we have just seen that 20 kilometers is 12 miles. So, 40 kilometers is 2 × 12 = 24 miles (approximately). Since these are largely estimates, one might also stretch the estimation where one seeks to see the equivalent of 20 miles in kilometers, by recognizing that the number 21 is a Fibonacci number and using that number to get the number of kilometers would get us to the number 34, which is a bit larger than expected but can be compensated accordingly as an estimation. Hopefully, these relationships between the U.S. system and the metric system serve to simplify your need to work more effectively with the two systems.

UNIT 48

The Origins of Today's Calendar

Our life today certainly demands knowing the day and the exact time, but thousands of years ago, the exact time, and even the exact day, may have been somewhat elusive and not as important. This is because the precise measurement of the Earth's movement about the Sun did not become very accurately known until almost a thousand years ago. As a result, the calendar has gone through many machinations throughout history.

The earliest formal calendars date to the Bronze age over 5000 years ago, a time when early civilizations built structures that

Figure 3.1

signaled the Sun's position at key moments of the year, such as a solstice or an equinox. In Wiltshire, England, on the mysterious Stonehenge monument (Figure 3.1), created ca. 3000 BCE, the Sun rises directly above the Heel Stone[4] at the beginning of the summer solstice.

The great Sphinx in Egypt (Figure 3.2) marks the spring or fall equinox when the Sun sets directly onto the right shoulder of the megalith looking at it from the front.

Early calendars were based either on the solar cycle or the lunar cycle, or a combination of both (lunisolar). Here we will consider the development of the civil calendar used today throughout the United States and Europe. One of the earliest precursors of our calendar is that of the Egyptian civilization during the Bronze Age. It consisted of a fixed 365 days and was principally a solar calendar as it was based on the year but focused on the first visible rising of Sirius, the brightest star in the heavens. This Egyptian calendar was almost a quarter of a day short of the actual solar year, and therefore every four years it fell behind by almost a day. Though attempts were made to correct this

---

[4]The Heel Stone is a large block of sarsen stone standing within the Avenue outside the entrance of the Stonehenge earthwork in Wiltshire, England.... It is 77.4 metres (254 ft) from the centre of Stonehenge circle. It leans towards the southwest nearly 27 degrees from the vertical.

Figure 3.2   Great Sphinx

error, they were resisted by Egyptian priests. It was not changed until 25 BCE when the first Roman Emperor, Augustus Caesar (63 BCE–14 CE) decreed that Egypt adopt the Julian calendar, which was proposed by the great general Julius Caesar (100 BCE–44 BCE) in 46 BCE.

Before the Julian calendar was adopted in Rome, the Roman calendar that existed was a lunar calendar introduced by Romulus, the legendary founder of Rome. Its year was divided into 10 months and contained 300 days, requiring his successor, Numa Pompilius (753 BCE–673 BCE), to add two months. This early calendar was in use for about 650 years, and since it was primarily based on the phases of the moon, it was shorter than the solar year. It needed constant corrections, which were applied during the month of February to preserve the seasons. Originally this lunar calendar started in March, at the time of the vernal equinox, which is the beginning of spring. The first few months were named for various gods, March being named for Mars, the god of war, because this was the month when military activity resumed after a hiatus at the end of the previous year. It is not clear how April was named, possibly from the Latin word *aperio* which means "to open," inspired by the opening of buds in spring. It is also speculated that April acquired its name from the goddess Aphrodite. Maia, the Earth goddess, was the inspiration for May. The Romans named June after Juno, the queen of gods and patroness of marriage. July drew its name from Julius Caesar, and August from his successor, Augustus Caesar. September, October, November and December were named based on the number of the month and were the seventh, eighth, ninth and tenth months, respectively.

When Julius took control of the Roman republic in 46 BCE, the seasons were occurring almost three months later than predicted by the calendar. Engaging the effort of Greek mathematicians and astronomers, who more accurately measured the solar year, he added January and February to the year extending it to 446 days changing the calendar to a solar calendar. January became the first month and was named after Janus, the Roman god of beginnings, while February's name is believed to stem from Februa, an ancient festival dedicated to the ritual springtime of cleaning and washing.

February in the Julian calendar originally had 23 days and was supposed to have a leap day of February 24 added every fourth year, but following Caesar's assassination, the priests misinterpreted the inclusion of the leap year and mistakenly added this day every three years. In order to bring the calendar back to its proper place, Augustus was obliged to suspend leap years for several decades. Even then, the calendar was based on a solar year of 365.25 days but the solar year, which varies because the earth's orbital speed, is not constant[5] and averages 365.24219 days, which means the Julian calendar gains a day every 128 years. Still the Julian calendar was the dominant calendar in the Roman empire and throughout Europe for more than 1600 years.

By the 16th century, the date of Easter had crept far away from the vernal equinox, which occurred around March 11 in 1582. During that year, Pope Gregory XIII promulgated a change in the calendar by eliminating 10 days and reducing the average year to 365.2425 days. This

---

[5]During the years 1900 to 2100 the shortest solar year in 2066 will be 365 days, 5 hours, 33 minutes and 54 seconds long. The longest solar year in 2093 will be 365 days, 6 hours, 1 minute and 6 seconds long.

corrected the inaccuracy which had developed, and it became known as the Gregorian calendar, which is the calendar we use today. Eventually most of the world adopted this corrected calendar, however, it took 170 years before Great Britain changed from the Julian calendar to the Gregorian calendar. In September 1752, 11 days needed to be omitted, so the day after September 2 was designated September 14. Figure 3.3 shows a page of the calendar for September 1752 taken from the almanac of Richard Saunders of Honiton, Gent, published in London, where you can see the deletion of the 11 days in September.

In some places, the Eastern Orthodox church still uses the Julian calendar, known as the Coptic calendar, and today its date is 13 days earlier than the corresponding Gregorian date. For example, the Gregorian date of January 13, 2030 will be January 1, 2030 in the Coptic calendar. Our Gregorian calendar still adds a day every four years, which would assume a solar year of 365.25 days, however, this is longer than the actual solar year, so it is necessary to eliminate the leap year every 100 years, except when a century year is divisible by

1752 September hath XIX Days this Year.

First Quarter, the 15th day at 2 afternoon.
Full Moon, the 23rd day at 1 afternoon.
Last Quarter, the 30th day at 2 afternoon.

M D	W D	Saints' Days Terms, &c	Moon South	Moon Sets	Full Sea at Lond.	Aspects and Weather
1	f	Day br. 3.35	3 A 27	8 A 29	5 A 1	□ ♃ ☿
2	g	London burn.	4 26	9 11	5 38	Lofty winds

According to an act of Parliament passed in the 24th year of his Majesty's reign and in the year of our Lord 1751, the Old Style ceases here and the New takes its place; and consequently the next Day, which in the old account would have been the 3d is now to be called the 14th; so that all the intermediate nominal days from the 2d to the 14th are omitted or rather annihilated this Year; and the Month contains no more than 19 days, as the title at the head expresses.

14	c	Clock slo. 5 m.	5 15	9 47	6 27	HOLY ROOD D.
15	f	Day 12 h. 30 m.	6 3	10 31	7 18	and hasty
16	g		6 57	11 23	8 16	showers
17	A	15 S. AFT. TRIN.	7 37	12 19	9 7	
18	b		8 26	Morn.	10 22	More warm
19	c	Nat. V. Mary	9 12	1 22	11 21	and dry
20	d	EMBER WEEK	9 59	2 24	Morn.	weather
21	e	ST. MATTHEW	10 43	3 37	0 17	☌ ♀ ☿
22	f	Burchan	11 28	☾ rise	1 6	□ ♃ ♀
23	g	EQUAL D. & N.	Morn.	6 A 13	1 52	☌ ☉ ♂
24	A	16 S. AFT. TRIN.	0 16	6 37	2 39	☌ ☉ ☿
25	b		1 5	7 39	3 14	
26	c	Day 11 h. 52 m.	1 57	8 39	3 48	Rain or hail
27	d	EMBER WEEK	2 56	8 18	4 23	☌ ♂ ☿
28	e	Lambert bp.	3 47	9 3	5 6	now abouts
29	f	ST. MICHAEL	4 44	9 59	5 55	⚹ ♄ ♀
30	g		5 43	11 2	6 58	

Figure 3.3

400, and then it is not designated a leap year. The years 1700, 1800, and 1900 were not leap years but the year 2000 was a leap year. So, anybody born on February 29 will not have a birthday in the year 2100 and will have to wait another 4 years to celebrate their birthday.

UNIT 49

How Eratosthenes Measured the Earth

Measuring the Earth today is not terribly difficult, but thousands of years ago, this was no mean feat. Remember, the word "geometry" is derived from "Earth measurement." Therefore, it is appropriate to consider this topic in one of its earliest forms. One of these measurements of the circumference of the Earth was made in about 230 BCE by the Greek mathematician Eratosthenes (276 BCE–194 BCE). His measurement was remarkably accurate as we will see when we review his amazing technique, which was based on the relationship of alternate-interior angles of parallel lines.

As the librarian of Alexandria, Egypt, Eratosthenes had access to records of calendar events. He discovered that at noon on a certain day of the year, in a town on the river Nile called Syene (now called Aswan), the sun was directly overhead because it was close to the equator. As a result, the bottom of a deep well was entirely lit and a vertical pole, being parallel to the rays hitting it, cast no shadow.

At the same time however, a vertical pole in the city of Alexandria, north of Syene, did cast a shadow. When that day arrived again, Eratosthenes measured the angle ($\angle 1$ in Figure 3.4) formed by such a pole and the ray of light from the sun going past the top of the pole to the far end of the shadow. He found it to be about $7°12'$, or $\frac{1}{50}$ of $360°$.

Assuming the rays of the sun to be parallel, he knew that the angle at the center of the Earth must be congruent to $\angle 1$ and hence must also measure approximately $\frac{1}{50}$ of $360°$. Since Syene and Alexandria were almost on the same meridian (a great circle arc, or longitude), Syene must be located on the radius of the circle passing through Syene and Alexandria, and the radius is parallel to the rays of the sun.

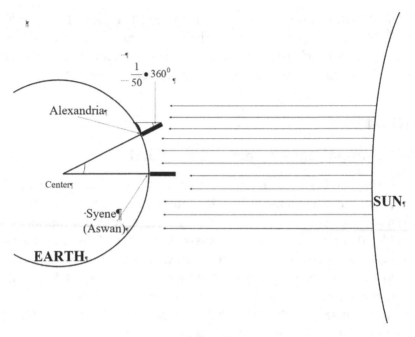

Figure 3.4

Eratosthenes thus deduced that the distance between Syene and Alexandria was $\frac{1}{50}$ of the circumference of the Earth.

It was now necessary to know the actual distance from Syene to Alexandria which was critical for the accuracy of the measurement. This distance was carefully surveyed yearly by professional bematists[6] and was used for taxation and agricultural purposes. It was calculated to be about 5,000 Greek *stadia*. A *stadion* was a unit of measurement equal to the length of an Olympic or Egyptian stadium and is estimated between 500 feet and 600 feet, but argued to be closer to 500 feet. Therefore, since Eratosthenes concluded that the circumference of the Earth was about 250,000 Greek stadia, which at 525 feet per stadium is about 24,858 miles! This is very close to modern calculation of the circumference of the Earth which is 24,901

---

[6] Bematists were specialists in ancient Greece who were trained to measure distances by counting their steps.

miles. When one considers how primitive measurement tools were at that time, we should be in awe of the accuracy that Eratosthenes achieved. It is also worth emphasizing that this knowledge was somewhat suppressed during the early Christian era until about 600 CE when most, if not all scholars accepted that the earth was a sphere.

UNIT 50

The Origin of the Cartesian Plane

High school geometry shows us that we can obtain a "picture" of a formula on a graph by plotting the values of the variables as coordinates of points on an *xy*-plane (Figure 3.5).

This significant achievement in mathematics, where algebra is applied to geometry, was the work of the French philosopher and mathematician René Descartes (1596–1650) and has become known

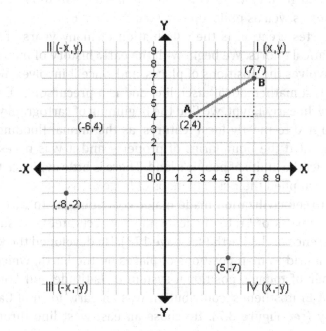

Figure 3.5   *xy*-plane

Figure 3.6   René Descartes

as Cartesian geometry. His contribution had a great impact on mathematicians as well as philosophers (see Figure 3.6).

Descartes creation is the culmination of many years of various mathematical efforts. We begin with the early history of map making which involves the positions of places, and since it involves measurement, it is a mathematical discipline and is a precursor of Cartesian geometry. In early Egypt, we find the beginning of cartography, which was confined to mostly local features, as the annual flooding of the Nile concealed the boundaries of property and it was necessary to reconstruct the limits of people's land. There is little evidence that the Egyptians applied their skill to maps of large areas.

The Greek civilization made major contributions to cartography first in the work of Eratosthenes who very accurately measured the circumference of the Earth (see page 173). He developed the key idea of using a grid to map positions of places on the Earth, which is the forerunner of today's positional system of Latitude and Longitude. Note that Eratosthenes' contribution was an early form of Cartesian geometry (see Figure 3.7). He chose an east-west line through the

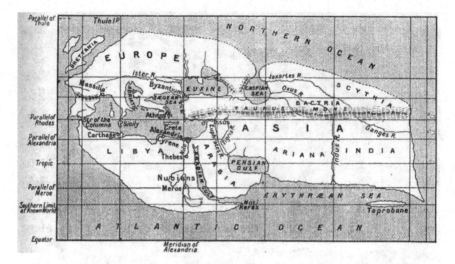

Figure 3.7   Eratosthenes' Map of the World

island of Rhodes, which is very close to 36 degrees north, to form one of the axes of the grid, with the other axis being a north–south line through Rhodes in Greece. This divided the known world into four roughly equal quadrants and seven lines were drawn parallel to each axis.

A most significant Greek contribution to Cartesian geometry was that of the astronomer and geographer Claudius Ptolemy (85 CE–165 CE), who about 140 CE wrote a seminal work *Guide to Geography* which included an effort to project the known spherical world onto a plane and assign coordinates to the major places, which was tantamount to latitude and longitude (see Figure 3.8).

Little progress was made for many years until the Arabic/Persian/Muslim world in the 9th century translated Ptolemy's *Geography* and made advances in cartography producing more accurate maps and which had a strong influence on the development of geography in Europe up to 1800. A world map produced by the Islamic mathematician Al-Khwarizmi (780–850) included the latitudes and longitudes of over 2000 places including cities, islands, mountains, seas, rivers and geographical regions.

Figure 3.8   Ptolemy's map of the known world

With the advent of the magnetic compass, maps for sailing ships, called *portolan* maps began to appear in the 14th century, and by the 15th century cartography blossomed in Europe around the middle of the century with the discoveries of the Portuguese explorers and the invention of the printing press.

The development of cartography was an important impetus for applying algebra to geometry and one of the earliest works to do this was that written in 1607 by the Croatian mathematician Marino Ghetaldi (1568–1626) when he produced a tract called *Variorum problematum collectio* containing 42 solutions of geometric problems. Soon after René Descartes published his great contribution to mathematics *La Géométrie*, written in 1637 as an appendix to *Discours de la méthode*. No doubt he was influenced by much of the work that preceded him, but he deserves much of the credit for his application of algebra to geometry which was in certain ways a unique concept and has earned him the name Cartesian geometry. This excerpt from Descartes' work highlights the effectiveness of his ideas:

> *In this way I believed that I could borrow all that was best both in geo-metrical analysis and in algebra and correct all the defects of the one by help of the other. The accurate observance of these few precepts gave me such ease in unravelling questions embraced in these two sci-ences, that not only did I reach solutions of questions I had formerly deemed exceedingly difficult, but even as regards questions of the solu-tion of which I continued ignorant, I was enabled to determine the means whereby, and the extent to which, a solution was possible....*

In the book *The Scientific Work of René Descartes: 1596–1650,*[7] J. F. Scott summarizes the importance of his contribution:

1. Algebra makes it possible to recognize the typical problems in geometry and to bring together problems which in geometrical dress would not appear to be related at all.

2. Algebra imports into geometry the most natural principles of divi-sion and the most natural hierarchy of method.

3. Not only can questions of solvability and geometrical possibility be decided elegantly, quickly and fully from the parallel algebra, without it they cannot be decided at all.

Florimond de Beaune (1601–1652) was a French jurist and ama-teur mathematician, who in 1649 published *Notes brièves* which was part of the first Latin edition of Descartes' *La Géométrie* and was the first important introduction to Cartesian geometry. Descartes' ideas no doubt revolutionized mathematics and had a profound effect on subse-quent research and discoveries. Today, it is a fundamental concept pres-ent in all of elementary mathematics and taught throughout the world.

---

[7] London—New York: Routledge, 1976.

UNIT 51

Why the Normal Bell-Shaped Curve is Called Normal

The "Normal" Distribution or so-called Bell-Shaped Curve, which one encounters in basic probability is shown in Figure 3.9 and resembles the cross-section of a bell. It is based on the following assumption about an infinite distribution of terms or numbers:

> *Each term or number is chosen or occurs in a completely random fashion with no bias or other influence as to its outcome.*

To help understand some of the properties of the Normal Curve, consider the following elementary example, the probability of whose outcomes are predicted by the Normal Curve. A perfectly balanced coin with an equal chance of turning up heads or tails is randomly tossed a very large number of times. The Normal Curve tells us the following probabilities about the number of heads and the number of tails that will occur:

1. The most probable outcome is that the number of heads equals the number of tails.

2. If the number of heads is greater than the number of tails by a certain amount, it is equally probable that the number of tails will exceed the number of heads by the same amount.

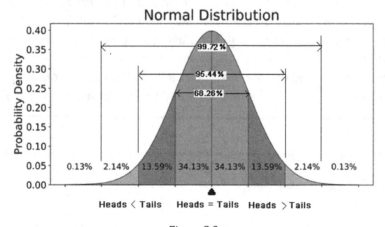

Figure 3.9

3. The greater the difference of the number of heads and tails, the less the chance that this difference will occur.

As an illustration, these are the probabilities out of 10,000 tosses of a perfectly balanced coin:

1. The most likely outcome is 5,000 heads and 5,000 tails.

2. The chance of tossing 6,000 heads and 4,000 tails is the same as the chance of tossing 6,000 tails and 4,000 heads.

3. The probability of tossing 7,000 heads and 3,000 tails is less than the probability of tossing 6,000 heads and 4,000 tails.

Figure 3.9 shows these probabilities as percentages within each section of the curve. The sections of the curve are based on what is called a *standard deviation*. The *standard deviation* is a measure that tells us the probability of how far a result will deviate from the middle, or mean, of the curve. In Figure 3.9, the total center shaded area on both sides of the mean which comprises 68.26%, contains all the terms that are one standard deviation or less from the mean. The total second shaded area on both sides of the mean which comprises 27.18%, contains all the terms that are between one and two standard deviations from the mean. The total third shaded area on both sides of the mean which comprises only 4.28%, contains all the terms that are between two and three standard deviations from the mean. Since the sum of these three areas is 99.72%, it is highly unlikely that results will be more than three standard deviations from the mean.

When we apply the Normal Curve to 10,000 tosses of a perfectly balanced coin, it turns out that the mean is 5,000, and the standard deviation calculates to 50. The total center shaded area on either side of the mean of the curve then tells us that the probability that the number of heads, or tails, will deviate from the mean of 5,000 by one standard deviation of 50, or less, is 68.26%. This means that the chance of getting between 4,950 and 5,050 heads (or tails) is 68.26%. The total of the first and second shaded areas on either side of the mean tells us that the probability that the results will deviate from the mean of 5000 by two standard deviations of 100 or less, is 95.44% The total of the three shaded areas on either side of the mean tells us that the probability that the results will deviate from the mean of 5000 by three standard deviations of 150 or less, is 99.72%. In other words, there is a very strong chance that out of 10,000 tosses, that almost all the results will be between 4850 heads and 5150 heads. The chance of a result being more than three standard deviations from the mean is very rare and is only 0.26%. This means that the chance of tossing the coin and getting more than 5150 heads or more than 5150 tails in 10,000 tosses is very unlikely.

The equation of the Normal Curve derived from these probabilistic considerations is somewhat advanced as it takes into account random factors. It is shown here only as an example of the complexity of the mathematics:

$$y = \frac{1}{\sigma \sqrt{2\pi}} e^{\frac{-(x-\mu)^2}{2\sigma^2}}.$$

In this formula μ (mu) is the mean or midpoint of the curve, π is the approximate constant 3.14159, e is the approximate constant 2.718282, and σ (sigma) is the *standard deviation*. The variable x can take on any value from minus infinity to plus infinity. These concepts are studied in a basic course in statistics or probability. At this point, though, let us take a look at why the term "normal" is used to describe this curve.

The assumption of complete randomness as the basis for this distribution, is very common for many natural events in the world. One that first comes to mind is the probability of giving birth to a boy or a girl. Assuming that the probability of either event occurring is equal, then the probability distribution of boys and girls will be a normal distribution just like the random tossing of a perfectly balanced coin. Similarly, assuming the distribution of many human traits to be totally random, such as height, weight, mental capacity, life span etc., then their distribution will approximate very closely that of a normal distribution. In fact, the distribution of many characteristics of any living organism, such as, in the case of a tree, its height, the number of its branches, the size of its leaves, the number of its blossoms, etc., can be described by a normal curve. And therein lies the reason for the curve's name, normal, because so many occurrences in the world are not subject to any constraints and happen in completely random ways. Curiously, the normal curve, though it deals with events that are completely random, turns out to be very smooth and symmetrical, and not skewed or appearing "abnormal" in any way.

UNIT 52

Measures of Central Tendency

In mathematics and in statistics we frequently use measures of central tendency — that is, we use various means such as the *arithmetic mean* (in common terms: the average), the *geometric mean*, and the *harmonic mean*. Our knowledge of them has been traced back to ancient times. As a matter of fact, the historian Iamblichos of Chalkis (ca. 250–330 CE) reported that Pythagoras (ca. 570–ca. 495 BCE), after a visit to Mesopotamia, brought back to his followers a knowledge of these three measures of central tendency. This may be one reason why today they are often referred to as *Pythagorean means.* We tend to use these means in statistical analyses, but there are some rather enlightening views when we inspect them and compare them geometrically.

Let's begin by introducing these three means for two values a and b as follows[8]:

- The *Arithmetic Mean is:* $AM\ (a, b) = a \Ⓐ\ b = \frac{a+b}{2}$,
- The *Geometric Mean is:* $GM\ (a, b) = a \Ⓖ\ b = \sqrt{a \cdot b}$, and
- The *Harmonic Mean is:* $HM\ (a, b) = a \Ⓗ\ b = \frac{2}{\frac{1}{a}+\frac{1}{b}} = \frac{2ab}{a+b}$.

---

[8]Arithmetic mean a and b is defined for all real numbers, the geometric mean is defined for $a \geq 0$, and $b \geq 0$, and the harmonic mean is defined for all real numbers a, $b > 0$ and $a + b \neq 0$.

The Arithmetic Mean

Before comparing the relative magnitude of these measures of central tendency or means, we ought to see what they actually represent. The arithmetic mean is simply the commonly-used "average" of the data being considered — that is, the sum of the data divided by the number of data items included in the sum. In a simple example, if we want to find the average — or arithmetic mean — between the two values of 30 and 60, we take their sum, 90, and divide it by 2 to get the arithmetic mean of 45.

We can also see the arithmetic mean as taking us to the middle of an arithmetic sequence — that is, a sequence with a common difference between terms — such as 2, 4, 6, 8, 10. To get the arithmetic mean, we divide the sum by how many numbers are being averaged: $\frac{2+4+6+8+10}{5} = \frac{30}{5} = 6$, which, as we expected, happens to be the middle number in the arithmetic sequence of values.

The Geometric Mean

Any time you have several factors, which are part of a product, and you want to find the "average" factor, this is the geometric mean. The average of interest rates is an application most used in everyday life.

Suppose you have an investment which earns 10% the first year, 40% the second year, and 50% the third year. What is its average rate of return? It is *not* the arithmetic mean, because what these numbers indicate is that during the first year, your investment was *multiplied* (not added to) by 0.10, in the second year it was multiplied by 0.40, and the third year it was multiplied by 0.50. The appropriate measure of central tendency is the geometric mean of these three numbers.

The geometric mean of two numbers a and b is $\sqrt{ab}$, and of three numbers a, b, and c is $\sqrt[3]{abc}$. Therefore, for the above situation the geometric mean of the 3 interest rates is $\sqrt[3]{(0.1)(0.4)(0.5)} \approx 0.27$.

The geometric mean gets its name from a simple interpretation in geometry. Consider the altitude to the hypotenuse of a right triangle shown in Figure 3.10, where *CD* is the altitude to the hypotenuse of right triangle *ABC*.

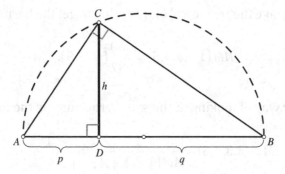

Figure 3.10

From the triangle similarity, $\triangle ADC \sim \triangle CDB$, we get the proportional relationship $\frac{AD}{CD} = \frac{CD}{BD}$, or $\frac{p}{h} = \frac{h}{q}$. This gives us $h = \sqrt{pq}$, which is one reason why h is considered the geometric mean between p and q.

The geometric mean is also the "middle" number of a geometric sequence. Take, for example, the geometric sequence, 2, 4, 8, 16, 32. To get the geometric mean of these five numbers, we find the fifth root of their product: $\sqrt[5]{2 \times 4 \times 8 \times 16 \times 32} = \sqrt[5]{32,768} = 8$, which is the middle number. As with the example of the arithmetic mean, the sequence has an odd number of numbers in order to have a middle number.

The Harmonic Mean

The harmonic mean is related very closely to the arithmetic mean in that if we take an arithmetic sequence such as 1, 2, 3, 4, 5, and take the reciprocals, we have what is called a harmonic sequence: $1, \frac{1}{2}, \frac{1}{3}, \frac{1}{4}, \frac{1}{5}$. We can tie the harmonic mean to the arithmetic mean by simply indicating that the harmonic mean is the reciprocal of the arithmetic mean of the reciprocals of the numbers. We would do this step-by-step, as follows:

To get the harmonic mean HM of a given sequence 1, 2, 3, 4, 5, we first find the arithmetic mean of the reciprocals of the sequence, that is,

$$AM\left(1, \frac{1}{2}, \frac{1}{3}, \frac{1}{4}, \frac{1}{5}\right) = \frac{1 + \frac{1}{2} + \frac{1}{3} + \frac{1}{4} + \frac{1}{5}}{5} = \frac{\frac{60+30+20+15+12}{60}}{5} = \frac{\frac{137}{60}}{5} = \frac{137}{300}(\approx 0.457).$$

We then take the reciprocal of this value to get the harmonic mean,

$$HM(1,2,3,4,5) = \frac{300}{137} (\approx 2.19).$$

Another way of looking at the same procedure is the following:

$$HM(1,2,3,4,5) = \frac{1}{AM\left(1, \frac{1}{2}, \frac{1}{3}, \frac{1}{4}, \frac{1}{5}\right)} = \frac{1}{1 + \frac{\frac{1}{2} + \frac{1}{3} + \frac{1}{4} + \frac{1}{5}}{5}}$$

$$= \frac{5}{1 + \frac{1}{2} + \frac{1}{3} + \frac{1}{4} + \frac{1}{5}} = \frac{300}{137} (\approx 2.19).$$

The harmonic mean is particularly useful in finding the average of rates over a common base. For example, consider the question of finding the average speed of a round trip journey. Suppose you were going at a rate of 30 mph and returning over the same route (the base) at 60 mph. One might be tempted to simply find the arithmetic mean, $\frac{30+60}{2} = 45$. This would be incorrect. Is it fair to give equal value to the 30 mph trip as to the 60 mph trip when the former took twice as long as the latter? Here we would invoke the harmonic mean, which would require us to get the reciprocal of the arithmetic mean of the reciprocals of the two numbers. That is, for the harmonic mean of 30 and 60, we get $\frac{1}{\frac{\frac{1}{30} + \frac{1}{60}}{2}} = \frac{2}{\frac{1}{30} + \frac{1}{60}} = \frac{2}{\frac{3}{60}} = \frac{120}{3} = 40$. This could, of course, be more simply done by using the formula $\frac{1}{\frac{\frac{1}{a} + \frac{1}{b}}{2}} = \frac{2ab}{a+b}$.

UNIT 53

Comparing The Three Means Algebraically

As a follow-up to our discussion of the 3 means presented in the previous unit, we present some unusual geometric methods for comparing the magnitude of the 3 means: Arithmetic, Geometric, and Harmonic. We shall show how these three means may be compared in size using simple algebra.

For two non-negative numbers a and b,

$$(a-b)^2 \geq 0$$
$$a^2 - 2ab + b^2 \geq 0$$

Add $4ab$ to both sides:
$$a^2 + 2ab + b^2 \geq 4ab$$
$$(a+b)^2 \geq 4ab$$

Take the positive square root of both sides:
$$a + b \geq 2\sqrt{ab}$$
$$\text{or} \quad \frac{a+b}{2} \geq \sqrt{ab}$$

This implies that the **arithmetic mean** of the two numbers a and b is greater than or equal to the **geometric mean**. (Equality is true, only if $a = b$.)

Similarly, we get our next result, which is a comparison of the geometric mean and the harmonic mean.

For the two non-negative numbers a and b,

$$(a - b)^2 \geq 0$$

$$a^2 - 2ab + b^2 \geq 0$$

Add $4ab$ to both sides:

$$a^2 + 2ab + b^2 \geq 4ab$$

$$(a + b)^2 \geq 4ab$$

Multiply both sides by ab:

$$ab\,(a + b)^2 \geq 4a^2b^2$$

Divide both sides by $(a+b)^2$:

$$ab \geq \frac{4a^2b^2}{(a+b)^2}$$

Take the positive square root of both sides:

$$\sqrt{ab} \geq \frac{2ab}{a+b}$$

This tells us that the **geometric mean** of the two numbers a and b is greater than or equal to the **harmonic mean**. (Here, equality holds whenever a or $b = 0$, or if $a = b$).

Therefore, $\frac{a+b}{2} \geq \sqrt{ab} \geq \frac{2ab}{a+b}$, which means the following: **Arithmetic mean ≥ Geometric mean ≥ Harmonic mean**.

Some may notice that the product of the arithmetic mean and the harmonic mean is $\frac{a+b}{2} \times \frac{2ab}{a+b} \times = ab$. Therefore, we can establish another relationship that ties the three means together:

$$AM(a, b) \times HM(a, b) = GM(a, b)^2, \text{ or written another way:}$$
$$(a \text{ⓖ} b)^2 = (a \text{Ⓐ} b) \times (a \text{Ⓗ} b).$$

With some further simple algebraic manipulation, we can get the following relationship:

$\frac{a}{\frac{a+b}{2}} = \frac{2a}{a+b} = \frac{2a \cdot b}{(a+b) \cdot b} = \frac{2ab}{(a+b)} \cdot \frac{1}{b} = \frac{\frac{2ab}{a+b}}{\frac{b}{1}}$, or in short, $\frac{a}{\frac{a+b}{2}} = \frac{\frac{2ab}{a+b}}{b}$, which then again shows us that $\frac{a}{AM(a,b)} = \frac{HM(a,b)}{b}$.

UNIT 54

Comparing the Three Means Geometrically

The comparison of the three means — the arithmetic, the geometric and the harmonic — in terms of their relative size was known to the

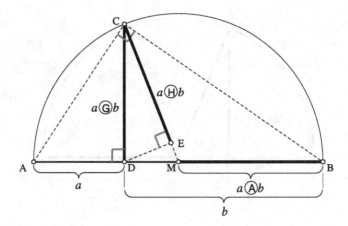

Figure 3.11

ancient Greeks, as we find in the writings of Pappus of Alexandria (ca. 250–350 CE). We will now embark on a geometric journey to consider various ways that the relative sizes of these means can be compared using simple geometric relationships.

In Figure 3.11, we have a right triangle with an altitude drawn to the hypotenuse, partitioning it into two segments of lengths a and b where $a \leq b$. Here we show the line segments that represent the three means and which allows us to compare their relative sizes. That is,

$$a \,\text{Ⓗ}\, b \leq a \,\text{Ⓖ}\, b \leq a \,\text{Ⓐ}\, b.$$

We explain below how these three segments are determined but to justify our visual observation, we begin by considering Figure 3.12, where we notice that CE is a leg of right triangle CED; therefore, $CE \leq CD$. Therefore, we have established that $a \,\text{Ⓗ}\, b \leq a \,\text{Ⓖ}\, b$.

Now to compare the geometric and the arithmetic means. Since the radius MC or MB of the circumscribed circle of triangle ABC is longer than the altitude to the hypotenuse of the triangle ABC, we have that $CD \leq MB$, which means that $a \,\text{Ⓖ}\, b \leq a \,\text{Ⓐ}\, b$. Combining these inequalities gives us that $CE \leq CD \leq MB$, or $a \,\text{Ⓗ}\, b \leq a \,\text{Ⓖ}\, b \leq a \,\text{Ⓐ}\, b$.

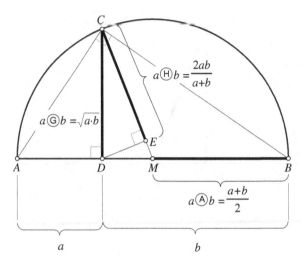

Figure 3.12

Now our task is to show that these three segments, whose relative lengths we have just established, actually represent the three means of a and b as we earlier stated.

First, we know that AB is the diameter of the circle with center at M, and radii $MA = MB = MC = \frac{a+b}{2}$, which is the arithmetic mean between a and b.

To find the geometric mean of a and b we begin with the altitude, CD, to the hypotenuse of right triangle ABC, which partitions the right triangle into two similar triangles ($\triangle ADC \sim \triangle CDB$) and therefore, $\frac{a}{CD} = \frac{CD}{b}$, which leads to $CD^2 = a \cdot b$; thus, $CD = \sqrt{a \cdot b}$, which is the geometric mean of a and b.

From the similar triangles ($\triangle CDM \sim \triangle ECD$) located in right triangle CDM, we can get $CD^2 = MC \cdot CE$, which yields $CE = \frac{CD^2}{CM} = \frac{a \cdot b}{\frac{a+b}{2}} = \frac{2ab}{a+b}$, which is the harmonic mean.

Having now justified that the line segments, $CE \leq CD \leq MB$, represent the various means of a and b, we have therefore, shown geometrically that $\frac{2ab}{a+b} \leq \sqrt{a \cdot b} \leq \frac{a+b}{2}$.

Some may notice that the product of the arithmetic mean and the harmonic mean is $\frac{a+b}{2} \cdot \frac{2ab}{a+b} = ab$. Therefore, we can establish another relationship that ties the three means together:

$AM(a, b) \cdot HM(a, b) = GM(a, b)^2$, or written another way:

$$(a \,\text{ⓖ}\, b)^2 = (a \,\text{ⓐ}\, b) \cdot (a \,\text{ⓗ}\, b).$$

Clearly, these means can be applied to more than two values. For example, when we find these means for three numbers, we have the following:

$$AM(a,b,c) = \frac{a+b+c}{3},$$

$$GM(a, b, c) = \sqrt[3]{a \cdot b \cdot c} \text{ and}$$

$$HM(a,b,c) = \frac{3}{\frac{1}{a}+\frac{1}{b}+\frac{1}{c}} = \frac{3abc}{ab+ac+bc}$$

We can extend these definitions to find the three means of any larger number values.

UNIT 55

The Rule of 72

There are topics in elementary mathematics that have surprisingly useful applications in the everyday world but unfortunately are not too often presented in the secondary school classroom. However, some experienced teachers do enhance their instruction with such applications. The topic we will unveil here will not only show off some of the beauty of mathematics, but also the power of its applications. We are going to present a method for determining how long it takes to double your money in the bank when it is compounded daily at a given annual rate. This is clearly good to know, but it is the unusual mathematical nature of this rule that we want to demonstrate here. So, read and enjoy it. It is called the "Rule of 72" because it is based on this number as you will soon discover.

The "Rule of 72" states that, roughly speaking *money will double in $\frac{72}{r}$ years when it is invested at an annual compounded interest* *rate* = *r*. So for example, if we invest money compounded daily at an annual interest rate of 8%, it will double its value in $\frac{72}{8}$ = 9 years. Similarly, if we leave our money in the bank at a daily-compounded rate of 6%, it will take $\frac{72}{6}$ = 12 years for this sum to double its value.

To investigate why this technique works and how accurate it is, we consider the compound interest formula: $A = P\left(1+\frac{r}{100}\right)^{n}$, where *P* is the principal invested for *n* interest periods, *r* is the annual interest rate, and *A* is the amount to which it grows. For the more advanced reader we shall investigate what happens when *A* = 2*P*.

The above equation then becomes:

$$2P = P\left(1+\frac{r}{100}\right)^{n} \quad \text{then } 2 = \left(1+\frac{r}{100}\right)^{n} \tag{1}$$

By taking logarithms of both sides and solving for *n*, we get:

$$n = \frac{\log 2}{\log\left(1+\frac{r}{100}\right)} \tag{2}$$

We then construct a table of values (Figure 3.13) from the formula comparing *r*, *n*, and *nr*.

r	n	nr
1	69.66071689	69.66071689
3	23.44977225	70.34931675
5	14.20669908	71.03349541
7	10.24476835	71.71337846
9	8.043231727	72.38908554
11	6.641884618	73.0607308
13	5.671417169	73.72842319
15	4.959484455	74.39226682

Figure 3.13

If we then take the arithmetic mean (the common average) of the *nr* values, we get 72.04092314, which is quite close to 72. Therefore, if you divide 72 by *r*, the result is close to *n*, the number of years it takes to double your money. Our "Rule of 72" quickly provides a very close estimate for doubling money at an annual interest rate of *r* % for *n* interest periods.

A curious reader, with a good knowledge of secondary school mathematics, might like to discover a "rule" for tripling or quadrupling money, similar to the way we dealt with the doubling of money. Taking the above equation (2), and modifying it for increasing your money *k times*, it becomes $n = \frac{\log k}{\log\left(1+\frac{r}{100}\right)}$, where *k* takes on integral values. When *r* = 8 %, this formula gives the value for *n* = 29.91884022 (log *k*). Now if take *k* = 3 (the tripling effect), *nr* = 239.3507218 log 3= 114.1993167. We could then say that for tripling money, we would have a "Rule of 114."

However far this topic is explored, the important issue here is that the common "Rule of 72" can be a useful tool to have at your disposal. If you should feel inspired to go further with this unit and extend it, that is admirable.

Chapter 4

Geometric Novelties

UNIT 56

The Platonic Solids

Platonic solids are regular convex polyhedra, which have congruent faces, equal length edges and equal polyhedral angles at each vertex. There are 5 such platonic solids, the tetrahedron, the octahedron, the icosahedron, the dodecahedron and the cube, which we show in Figure 4.1. The famous Greek philosopher Plato (ca. 450 BCE) is credited among the early writers to have alluded to the solids around 300 BCE. In his work entitled *Timaeus,* he related the cube, octahedron, icosahedron, and tetrahedron, respectively, to earth, air, water and fire. The dodecahedron Plato claimed was used by God to arrange the constellations in heaven. He chose these because of their physical structure as he related them to these aspects of life.

Their initial entry into the mathematical realm seems to have been in Euclid's book, *Elements.* There he also describes their construction and indicates that there are no other such solids with this kind of regularity. There are number of ways in which we can prove that there are only five regular polyhedra. The proof that Euclid uses in the *Elements* goes as follows:

The solids are comprised of vertices, faces and edges and each has its limits. The vertices are comprised of at least 3 faces. However, the plain angle of each face at a vertex must be less than 120°, since if they were each 120°, then the sum of 3 such would be 360° and there would be no three-dimensional vertex.

As we look at regular polygons, those with 6 or more sides have angles exceeding 120°, and therefore are not eligible to be faces of a regular polyhedron. Thus, we will consider faces that are regular polygons of 3, 4, and 5 sides, namely, triangular faces, square faces, and pentagonal faces. The regular triangle (equilateral triangle) has vertices of 60°. Placing them to form a three-dimensional vertex, commonly known as a polyhedral angle, we find that there are 3 possibilities, and they generate the tetrahedron, octahedron, and icosahedron. When we consider faces of the square, where the vertex angle is 90°, the only possibility for creating a polyhedral angle with face angles of 90° is that which forms a cube. Now considering a possible face of a regular polygon which is a pentagon, each vertex angle is 108° which again allows us only 3 faces at the polyhedral vertex, leading us to construct a dodecahedron. When we count the number of regular polyhedra thus constructed we end up with 5, which we show in Figure 4.1.

Polyhedron		Vertices	Edges	Faces	Number of edges of each face	Number of faces at each vertex
tetrahedron		4	6	4	3	3
octahedron		6	12	8	3	4
icosahedron		12	30	20	3	5
dodecahedron		20	30	12	5	3
cube		8	12	6	4	3

Figure 4.1   The five platonic solids

Possible values for n	Possible values for m
3	3
4	3
3	4
5	3
3	5

Figure 4.2

The famous Swiss mathematician Leonhard Euler (1707–1783) proved a very important relationship that applies to all polyhedra. Given that V equals the number of vertices, E equals the number of edges, and F equals the number of faces it follows that $V - E + F = 2$. When we apply this to a tetrahedron, we see that $4 - 6 + 4 = 2$.

We offer a proof of the existence of exactly 5 regular polyhedra which uses simple algebra. We begin by letting n equal the number of edges of each face, and m equal the number of edges at each vertex. Therefore, $nF = 2E = mV$. This then gives us: $F = \frac{2E}{n}$, and $V = \frac{2E}{m}$. Substituting for the values of F and V in the above Euler equation we get: $\frac{2E}{m} - E + \frac{2E}{n} = 2$. By dividing both sides of this equation by E, we get: $\frac{2}{m} - 1 + \frac{2}{n} = \frac{2}{E}$.

By dividing both sides of the equation by 2, we get: $\frac{1}{m} - \frac{1}{2} + \frac{1}{n} = \frac{1}{E}$. Rearranging the terms, gives us: $\frac{1}{m} + \frac{1}{n} = \frac{1}{2} + \frac{1}{E}$. Since, $E > 0$, we can easily establish the following relationship, $\frac{1}{m} + \frac{1}{n} > \frac{1}{2}$. We can then show that there are only 5 relationships, for m and n, as shown in Figure 4.2.

You now have a sense of the Platonic solids or as we now can call them regular polyhedra and a proof that there are only 5 of these lovely solids that can exist.

Unit 57

Euler's Theorem Revisited

In the previous unit, we introduce Euler's interesting theorem establishing the relationship between the number of vertices, edges, and faces of polyhedra as follows:

Vertices – Edges + Faces = 2

This relationship is shown for the five platonic solids but it is also true for all polyhedra such as the double square pyramid shown in Figure 4.3 and all the polyhedra shown in Figure 4.4.

There are many proofs of Euler's theorem, which are somewhat advanced and can be found at the website: https://www.ics.uci.edu/~eppstein/junkyard/euler/.

What is fascinating about this theorem is that it can also be found in certain natural phenomena. Let us examine two such examples. The

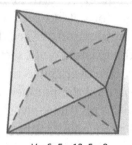

V = 6, E = 12, F = 8
6 – 12 + 8 = 2

Figure 4.3

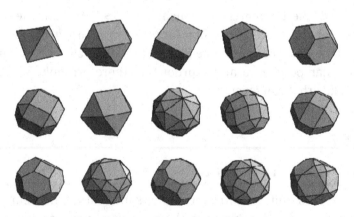

Figure 4.4   Various Polyhedra

first is a topographical application, which applies to a mountain range shown in Figure 4.5. The features of a mountain range can be identified as three types:

1. A Peak which is the top of a mountain.
2. A Pass which is a dip between two mountain tops.
3. A Pit which is valley that leads to a mountain pass.

The relationship between these three mountain related features is as follows:

$$\text{Peaks} - \text{Passes} + \text{Pits} = 2$$

This is an application of Euler's polyhedron theorem where Peaks are analogous to Vertices, Passes are analogous to Edges and Pits are analogous to Faces. An example of this relationship can be seen in the

Figure 4.5

simple configuration of two peaks connected by one pass and surrounded by one pit: Peaks − Passes + Pits = 2 −1 + 1 = 2.

If one adds a third peak creating two passes but still surrounded by one pit, we have another example: Peaks − Passes + Pits = 3 − 2 + 1 = 2.

For every Peak that one adds, a pass is created preserving the relationship, which continues to be true no matter how complex the mountain range is, such as that shown in Figure 4.6.

Another fascinating application of Euler's theorem is to a weather map of Highs, Lows and the boundary between two Highs or two Lows known as a Col. The Col is a region of slightly elevated pressure, which separates the two systems. The relationship between these weather systems is: Highs − Cols + Lows = 2.

Figure 4.6

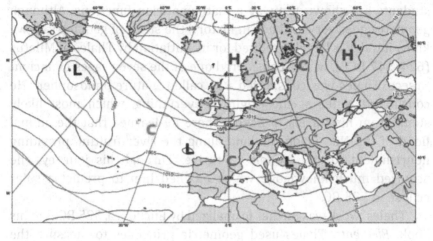

Figure 4.7

The Highs are analogous to Peaks, the Cols to Passes and the Lows to Pits. In the weather map in Figure 4.7, which shows these three systems, the relationship yields:

$$\text{Highs} - \text{Cols} + \text{Lows} = 2 - 3 + 3 = 2$$

If a High were to split into two Highs, or a Low were to be split into two Lows, a Col would be formed preserving the relationship. So, the next time you look at a mountain range, or a weather map, you can appreciate the simple mathematical relationship that applies to these natural phenomena.

UNIT 58

The Beginnings of Geometry

The United States secondary school curriculum is one of the few in the world, where an entire year is devoted to the study of geometry in a formal sense, and where theorems are proved before being accepted. It is the one time in the secondary school curriculum that students are shown the way a mathematician functions. Although geometry goes very far back in history, the study of geometry as we know it today is largely credited for its initiation by Thales of Miletus (624–547 BCE), who is considered one of the seven sages of ancient Greece. Thales was considered a so called "nature philosopher." He considered water as the origin of everything, which most likely stems from his travels to Egypt. There he learned that the annual flood of the Nile made the land along the river fruitful, providing nourishment for the people to thrive. Through his journeys, he acquired much knowledge which enabled him to predict a solar eclipse in 585 BC.

Thales' work was further formalized by Euclid (ca. 325 BCE) in his book, *Elements*. Thales used geometric principles to measure the height of pyramids and the distances of ships. Yet, he is revered today for some of the very basic relationships in Euclidean geometry. The theorem that bears his name is the one where he states that for a triangle inscribed in a circle, where one side of the triangle is the diameter of the circle, the angle opposite that side is a right angle. We show Thales theorem in Figure 4.8, where AC is the diameter and angle ABC is a right angle.

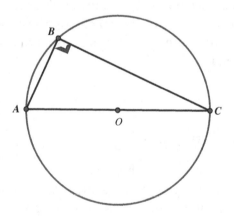

Figure 4.8

Thales was also the first to establish that two vertical angles are equal. In other words, in Figure 4.9, where lines *AB* and *CD* intersect at point *P*, angle *APC* is equal to angle *BPD*.

Perhaps on a more sophisticated level, Thales also determined that parallel lines cut off proportional segments on intersected lines. Symbolically, we can see that in Figure 4.10, where parallel lines *AB* and *CD* are intersected by lines *EAC* and *EBD* we have the following relationship: $\frac{EA}{EC} = \frac{EB}{ED} = \frac{AB}{CD}$.

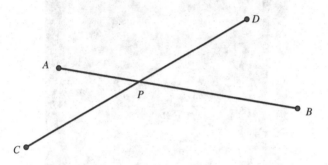

Figure 4.9

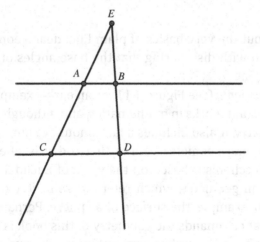

Figure 4.10

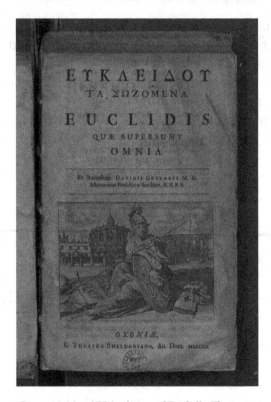

Figure 4.11   1704 edition of Euclid's *Elements*

To round out the very basics of plain Euclidean geometry, Thales is also credited with discovering that the base angles of an isosceles triangle are equal.

Euclid's *Elements* (see Figure 4.11 for an early exemplar) is one of the most important works in mathematics, and although it is largely a study of geometry, it also includes a fair amount of number theory.

The traditional geometry course that today is offered in most American high schools is based on the work of Euclid. It is therefore called Euclidean geometry, which refers to geometry on a plane as opposed to, for example, the surface of a sphere. Perhaps the significant aspect that commands the geometry of this book is Euclid's fifth postulate, which reads: "if a line segment intersects two straight lines, forming two interior angles on the same side of that given line, such that the sum of their measures is less than two right angles, then the two lines, if extended indefinitely, will meet on that same side of the given line, where those two angles have a sum less than two right angles." This was vastly simplified in 1846 by the Scottish mathematician John Playfair (1748–1819), who stated an equivalent postulate, known today as Playfair's axiom and reads: "In a plane, given a line and a point not on it, at most one line parallel to the given line can be drawn through the point." It is this axiom that today guides the basics in Euclidean geometry.

The *Elements* had also great influence beyond mathematics. Amazingly, President Abraham Lincoln in his autobiography of 1860 stated "After he was twenty-three and had separated from his father, he studied English grammar — imperfectly, of course, but so as to speak and write as well as he now does. He studied and nearly mastered the six books of Euclid, since he was a member of Congress." Although the first six books largely do geometry, they provided for him an ability to improve his facilities and especially his powers of logic and language. He even referred to Euclid in the famous fourth debate he had in 1858 with Judge Stephen A. Douglas (1813–1861) in Charleston, South Carolina. He said: "If you have ever studied geometry, you remember that by a course of reasoning, Euclid proves that the sum of all the angles of a triangle is equal to two right angles. Euclid has shown you how to work it out. Now, if you undertake to disprove that proposition, and to show that it is erroneous, would you prove it to be false by calling Euclid a liar?" It was also known at the time that when Lincoln traveled by horseback, he always carried a copy of Euclid's *Elements* in his saddlebags. Although Lincoln had no formal education, we can see that his devotion to learning was truly remarkable.

Although we have very little information about Euclid's biography, we can see the legacy that he initiated through his famous book *Elements,* and how to the present day it is the basis, not only for our high school geometry course, but it also had a role in our logical thinking, as it exhibited Abraham Lincoln's application of Euclid's reasoning. Perhaps more importantly, the *Elements* served to influence many major scientists such as Nicolaus Copernicus, Galileo Galilee, Johannes Kepler, and Isaac Newton among others.

UNIT 59

The Origin of the American Secondary School Geometry Course

The previous unit gives you some historical perspective of where formal geometry has its roots. The question is then asked, how did the United States high school geometry curriculum evolve?

As we said earlier, the United States education system is one of the few in the world where students study one year of geometry while still in high school. The curriculum for this course is ultimately based on Euclid's *Elements.* Yet the path the high school geometry curriculum took is rather interesting. We could begin with the Scottish mathematician Robert Simson (1687–1768) (Figure 4.12), who set out to prepare a perfect text of Euclid's first 6 books together with the 11th and 12th books, and first published it as a complete book in Glasgow, Scotland in 1756. We show the title page of the 1787 edition in Figure 4.13.

This work was then adopted by the French mathematician Adrien-Marie Legendre (1752–1833), who wrote a textbook in 1794

Figure 4.12   Robert Simson

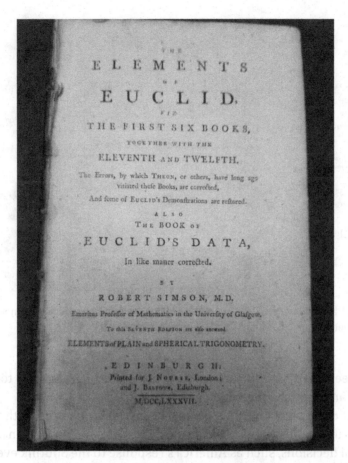

Figure 4.13   Euclid's elements

entitled *Eléments de géométrie,* which in turn became the model for the American basis for the high school geometry course as we know it today. Legendre's book was translated from French by David Brewster in 1828 and was later adapted by the American mathematician Charles Davies (1798–1876) and entitled *Elements of Geometry* and a second book *Elements of Geometry and Trigonometry,* (Figure 4.14) in 1862 and 1872, respectively. This then became the model for teaching geometry in the United States. Naturally, there

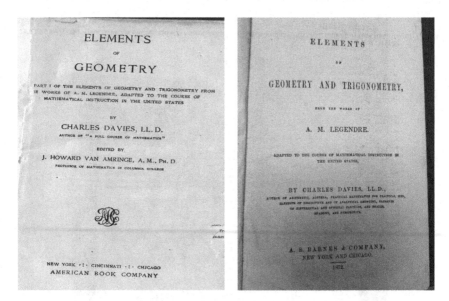

Figure 4.14   Davies' American adaptation of Simson's and Legendre's geometry and trigonometry

have been many modifications over the years till we get to today's American high school geometry course.

It is quite clear that this geometry curriculum has gone through numerous modifications over the past decades — often influenced by political decisions, such as America's response to the Sputnik event of October 4th 1957, that forced a strong review by the School Mathematics Study Group (1958–1977) consisting of math teachers and mathematicians who redefined what and how mathematics should be taught. The result was called "The New Math." This was followed by the "Back to Basics" movement, after which technology dominated the curriculum platform. Through these modifications, the geometry course has endured and remains a major part of the American high school curriculum. Therefore, the high school geometry course seems to have an influence well beyond mathematics in that it trained students to do logical thinking.

UNIT 60
The Pythagorean Theorem and Its Origins

The one topic that most people recall from their high school mathematics studies is the Pythagorean theorem. Although the relationship was already known before Pythagoras (as you will see below), it is appropriate that the theorem should be named for him, since Pythagoras (or one of the Pythagoreans) was the first to give a proof of the theory — at least as far as we know. Historians suppose that he used the squares as shown in Figures 4.15 and 4.16 — perhaps inspired by a pattern of floor tiles. Here is a brief version of the proof.

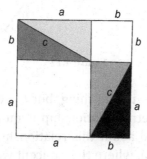

Figure 4.15

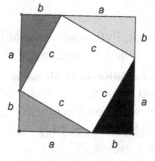

Figure 4.16

To show that $a^2 + b^2 = c^2$, you need only subtract the four right triangles, with sides a, b, and c from each of the two larger squares, so that in Figure 4.15 you end up with $a^2 + b^2$, and in Figure 4.16 you end up with c^2. Therefore, since the two original squares were the same size and we subtracted equal quantities from each, we can conclude that $a^2 + b^2 = c^2$, which is shown in Figure 4.17 with the two figures of the same area.

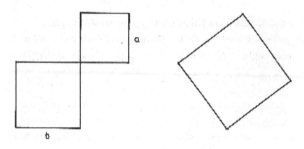

Figure 4.17

To prove a theorem is one thing, but to come up with the idea establishing this geometric relationship is quite another. It is likely that Pythagoras learned about this relationship on his study trip to Egypt and Mesopotamia, where this concept was known and used in construction projects.

Egypt

During his travels to Egypt, Pythagoras probably witnessed the measuring method of the so-called Harpedonapts (rope stretchers). They used ropes tied with 12 equidistant knots to create a triangle with two sides of length 3 and 4 units and a third side of 5 units. Since the converse of the Pythagorean theorem is also true, that is if $a^2 + b^2 = c^2$, then the triangle is a right triangle, they were able to "construct" a right angle, as shown in Figure 4.18.

They applied this knowledge to survey the banks of the River Nile after the annual floods in order to rebuild rectangular fields for the farmers. They also employed this method in laying the foundation

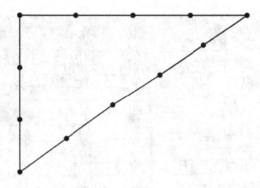

Figure 4.18

stones of temples. To the best of our knowledge, the Egyptians did not know of the generalized relationship given to us by the Pythagorean theorem. They seem to have only known about the special case of the triangle with side-lengths 3, 4 and 5, which produced a right triangle. This perhaps was arrived at by experience and not by some sort of formal proof.

Mesopotamia

In Mesopotamia, mathematicians were even able to produce further triples of numbers, fulfilling the Pythagorean condition of $a^2 + b^2 = c^2$, as we can see on a Babylonian clay tablet from ca. 1800 BCE, known as the Plimpton 322[1] (see Figure 4.19). The tablet was part of a collection of about a half-million of such tablets found in the mid-19th century in Mesopotamian digs, of which about 300 were identified as having mathematical significance. The tablet is written in Old Babylonian (or cuneiform) script, which is in the sexagesimal system (base 60). It shows us the high level of mathematical knowledge that existed well before the Greeks.

The transcript of these cuneiforms as shown in the two shaded columns of Figure 4.20 (in our base 10 system) gives a strong

---

[1]This tablet is in the permanent collection of the Columbia University (New York) Library.

Figure 4.19

a	b	c	m	n
120	119	169	12	5
3456	3367	4825	64	27
4800	4601	6649	75	32
13500	12709	18541	125	54
72	65	97	9	4
360	319	481	20	9
2700	2291	3541	54	25
960	799	1249	32	15
600	481	769	25	12
6480	4961	8161	81	40
60	45	75	2	1
2400	1679	229	48	25
240	161	289	15	8
2700	1771	3229	50	27
90	56	106	9	5

Figure 4.20

indication of their knowledge of the Pythagorean triples. These two columns list the leg and hypotenuse of several Pythagorean triples.

Here, we notice that the three left-numbers in each row satisfy the Pythagorean theorem, $a^2 + b^2 = c^2$, which we refer to as Pythagorean triples, while the two numbers on the right side (labeled m and n) are the numbers that can be used to generate these triples as follows: $a = 2mn$, $b = m^2 - n^2$, and $c = m^2 + n^2$.

Pythagorean triples have also been discovered in Northern Europe in megalithic rings, where they are displayed as triples of numbers which are, in large measure, accurate Pythagorean triples. However, in Babylonia we not only find Pythagorean triples, but we also find problems which can only be solved with a proper knowledge of the Pythagorean theorem.

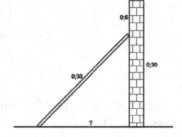

You *make* 0;30 *hold itself*, you see 0;15.
You *tear out* 0;6 from 0;30, you see 0;24.
You *make* 0;24 *hold itself*, you see 0;9,36.
You *tear out* 0;9,36 from 0;15, you see 0;5,24.
0;5,24 has what *square side*? It has 0;18 as *square side*.
It has moved away 0;18 (*nindānu*) on the ground.

Figure 4.21

The Babylonians derived the *Pole against the wall problem.* As shown in Figure 4.21, if a pole of length 0;30 units slips down 0;6 units along the 0;30-unit wall, how far is the base of the pole from the base of the wall?[2] (see the unit Ancient Number Systems, page 1).

This calculation is seen geometrically in Figure 4.22.

The calculation using our words and our number symbols would translate to the following:

Square 0.5, you will get 0.25.
Subtract 0.1 from 0.5, you will get 0.4.
Square 0.4, you will get 0.16.
Subtract 0.16 from 0.25, you will get 0.09.
0.09 is the area of a square, so its side is 0.3.

---

[2] The Babylonians used a number system with the base 60 (instead of 10 as we use to do). In particular, 0;30 means $0+\frac{30}{60}=0.5$, 0;6 means $0+\frac{6}{60}=0.1$, and 0;9,36 means $0+\frac{9}{60}+\frac{36}{60^2}$ etc.

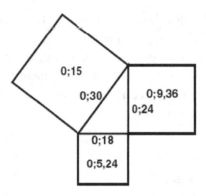

Figure 4.22

The stick has slipped by 0.3 which, when converted to the sexagesimal system, is 0;18.[3]

On another clay tablet, YBC 7289 (Yale University Babylonian Collection), we can see that the Babylonians already applied the Pythagorean theorem to calculate a rather accurate approximation of $\sqrt{2}$. In Figure 4.23, there are three pictures of this tablet; the first shows the original tablet, the second displays it with accentuated marking, and the third shows the values of lengths of its sides. On this tablet there is a square, whose sides have length 1, and with a value written along the diagonal. There we have an isosceles right triangle — half the square — where by using the Pythagorean theorem, we can determine that this isosceles right triangle with legs of length 1, has the hypotenuse length $\sqrt{2}$, since $1^2 + 1^2 = 2 = c^2$, so that $c = \sqrt{2}$.

The numbers along the diagonal are: 1; 24, 51 and 10 which means: $1 + \frac{24}{60} + \frac{51}{60^2} + \frac{10}{60^3} = 1.414212963$. This is a very close approximation of the value of $\sqrt{2} = 1.4142135\ldots$. The second line, 42; 25, 35, is the product of this number, $\sqrt{2}$, with the length of the (upper left) leg, given as 30. That is, the second line's value is in base 60, which is:

$$42 + \frac{25}{60} + \frac{35}{60^2} = 42.42638889, \quad \text{while} \quad 30 \times \sqrt{2} = 42.42640687\ldots,$$ a very close approximation!

---

[3] $0.3 = \frac{18}{60}$, which we write as 0;18.

Figure 4.23

India

We assume — because of geographical reasons — that Pythagoras learned about the relationship involving right triangles in Egypt or Mesopotamia. Some historians, however, argue that he may also have learned about it in India. Similarities were discovered between Indian philosophies and Pythagoras' principles. In the Sulva Sutra ("rules of the rope") of Baudhayana (about 800 BCE), we can already find problems which refer to the Pythagorean relationship:

> *"A rope stretched along the length of the diagonal produces an area, which the vertical and the horizontal sides make together."*[4]

---

[4]This means that if the side of a square is the length of the diagonal of a rectangle, then it has an area that equals the sum of the areas of the squares whose side-lengths are the sides of the rectangle.

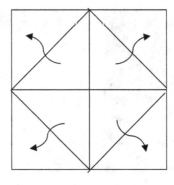

Figure 4.24

And for the special case of an isosceles right triangle there is:

"A chord which is stretched across the square produces an area double the size of the original square." [5]

As a special case of the previous relationship, which was applied to an isosceles right triangle, we can observe that the folding and unfolding (shown in Figure 4.24) is analogous to the argument that Socrates uses in his style of "teaching" Meno's slave in Plato's *Menon Dialogue* (380 BCE), by taking a concrete example in a very simple situation.

These are the earliest recorded statements of the Pythagorean theorem in Indian mathematical literature. Other books written in subsequent centuries contain similar statements, therefore we can assume that Baudhayana's Sulva Sutra was well known in this region. One could further assume that Pythagoras may have become familiar with this idea when he came in contact with this culture.

China

Despite a historical emphasis on Western culture, one must also consider what happened in the Far East. Was Pythagoras the first to

---

[5] This recognizes that the diagonal of a unit square must be equal to $\sqrt{2}$, since that length — as the side of a square — would produce an area of 2, or twice that of the original square.

prove the theorem that bears his name or was he anticipated in the Far East? In the book *Chou Pei Suan Ching*, one of the oldest Chinese books on mathematics (perhaps even the oldest), we can also find the Pythagorean theorem. However, it is not clear when it was written. Some historians believe it was in the 12th century BCE, while others place the work in the first century BCE. So, it is not clear whether the Indian mathematicians learned the Pythagorean theorem from the Chinese (or vice versa) or whether each discovered the theorem independently. The last assumption seems most likely, because in Chinese mathematics, the Pythagorean Theorem was more of an exercise in arithmetic, whereas in Indian or Babylonian mathematics, the principles of the Pythagorean Theorem arose from geometrical measurement.

Almost as old as the *Chou Pei Suan Ching* is a book by Chui-Chang Suan-Shu (or Jiuzhang Suanshu), *Nine Chapters on the Mathematical Arts*, which is probably the most influential of all ancient Chinese books on mathematics. It contains a collection of 246 problems on surveying, engineering and other subjects that relate to mathematics. The Pythagorean theorem is called Gougu (or Kouku). "Gou" means the shorter leg of a right triangle (originally: of a carpenter's square) and "gu" refers to the longer leg. The hypotenuse is called "hsien," which translates to "spanned chord."

One of the problems is that of the "Broken Bamboo." In Figure 4.25 we see that a bamboo pole is 10 feet tall. The upper end is broken, and the top reaches the ground 3 feet from the stem. The question asked is to find the height of the break. Such is the ubiquity of this famous theorem.

In contrast to other cultures which did not provide justifications for the Pythagorean Theorem, the Chinese gave a "proof," at least for the special case of the 3-4-5-triangle.

If we subtract the four dark right triangles in both squares of Figure 4.26, then we will be left with c^2 in the first square and $a^2 + b^2$ in the second square.

Regardless of where it originated, the Pythagorean theorem has fascinated mathematicians and amateurs alike through the ages. Famous people such as Plato, Euclid, Aristotle, Leonardo da Vinci, and

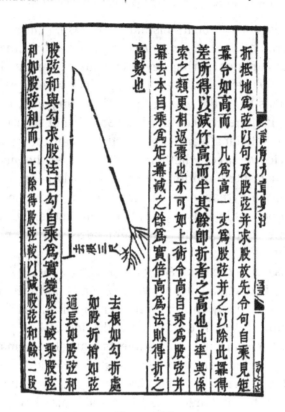

Figure 4.25

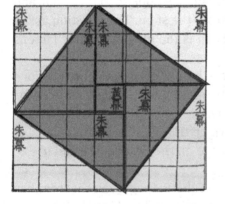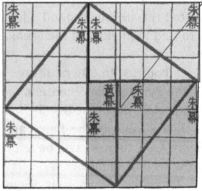

Figure 4.26

U.S. President James A. Garfield produced original proofs of this time-honored theorem. Mathematics enthusiasts have searched for proofs of this famous theorem for millennia. In fact, in 1940, Elisha Scott Loomis (1852–1940), the professor emeritus of mathematics at Baldwin-Wallace College, published the second edition of his *The Pythagorean Proposition*, which contains 367 different proofs of the Pythagorean Theorem. Since then, there have been many other proofs produced and proudly published in mathematics journals. Yet, as Loomis clearly noted, not one of the proofs of the Pythagorean Theorem that he presented in his book used trigonometry since, as he correctly pointed out, the field of trigonometry depends on the Pythagorean theorem for its primary relationship, namely, that *sin²A + cos²A* = 1. Therefore, using trigonometry to prove the Pythagorean theorem would be tantamount to circular reasoning, since proving a theorem with a "tool" that depends on the theorem is not proper logic. The establishment of *sin²A + cos²A* = 1 can be easily shown by applying the Pythagorean theorem to the sine and cosine ratios for right triangle *ABC* shown in Figure 4.27.

For right triangle *ABC* (Figure 4.27), we know that the sine ratio is: $\sin A = \frac{a}{c}$. and the cosine ratio is: $\cos A = \frac{b}{c}$. As we said, the basic trigonometric relationship that underpins the field of trigonometry is: $\sin^2 A + \cos^2 A = 1$. From the above defining ratios we get $\sin^2 A + \cos^2 A = \left(\frac{a}{c}\right)^2 + \left(\frac{b}{c}\right)^2 = \frac{a^2}{c^2} + \frac{b^2}{c^2} = \frac{a^2 + b^2}{c^2}$. However, from the Pythagorean theorem we have $a^2 + b^2 + c^2$. Therefore, $\frac{a^2 + b^2}{c^2} = 1$, and then $\sin^2 A + \cos^2 A = 1$.

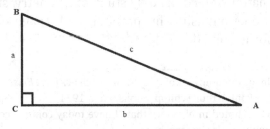

Figure 4.27

To further appreciate this revelation, consider the words of the great scientist, Albert Einstein[6]:

> *"I remember that an uncle told me the Pythagorean theorem before the holy geometry booklet had come into my hands. After much effort, I succeeded in "proving" this theorem on the basis of the similarity of triangles ... for anyone who experiences [these feelings] for the first time, it is marvelous enough that man is capable at all to reach such a degree of certainty and purity in pure thinking as the Greeks showed us for the first time to be possible in geometry."*

With this historical background of this famous theorem, one gets a much greater appreciation for its role in the foundation of geometry.

UNIT 61

President James A. Garfield's Contribution to Mathematics

What do the following three men have in common: Pythagoras, Euclid and James A. Garfield (1831–1881), the twentieth president of the United States?

After some moments of perplexity, we shall relieve your frustration by telling you that all three fellows proved the Pythagorean theorem. The first two fellows should not be a surprise, but President Garfield? He was not a mathematician. He did not even study mathematics. As a matter of fact, his only study of geometry, started some 25 years before he published his proof of the Pythagorean theorem, and it was informal and done alone.[7]

---

[6] From pp. 9–11 in the opening autobiographical sketch of *Albert Einstein: Philosopher-Scientist*, edited by Paul Arthur Schilpp, published in 1951.

[7] In October 1851 he noted in his diary that "I have today commenced the study of geometry alone without class or teacher."

While being a member of the House of Representatives, Garfield, who enjoyed "playing" with elementary mathematics, stumbled upon a cute proof of this famous theorem. It was subsequently published in the *New England Journal of Education* after being encouraged by two professors (Quimby and Parker) at Dartmouth College, where he went to give a lecture on March 7, 1876. The text of the two professors begins with

"In a personal interview with General James A. Garfield, Member of Congress from Ohio, we were shown the following demonstration of the pons asinorum,[8] which he had hit upon in some mathematical amusements and discussions with other Members of congress. We do not remember to have seen it before, and we think it something on which the members of both houses can unite without distinction of party."

Garfield's proof is actually quite simple, and therefore can be considered "beautiful." Beauty in mathematics is often in simplicity! We begin the proof by placing two congruent right triangles ($\triangle ABE \cong \triangle DCE$) so that points B, C and E are collinear as shown in Figure 4.28, and as a result, a trapezoid, $ABCD$, is formed. Notice also that since $\angle AEB + \angle CED = 90°$, $\angle AED = 90°$, thus, making $\triangle AED$ a right triangle.

$$\text{The Area of trapezoid } ABCD = \frac{1}{2}(\text{sum of bases})(\text{altitude})$$

$$= \frac{1}{2}(a+b)(a+b)$$

$$= \frac{1}{2}a^2 + ab + \frac{1}{2}b^2$$

---

[8]This would appear to be a wrong reference, since we usually consider the proof that the base angles of an isosceles triangle are congruent as the pons asinorum, or "bridge of fools."

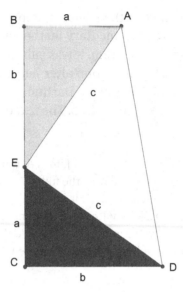

Figure 4.28

The sum of the three triangles, which just happens to also be the area of the trapezoid) is:

$$= \frac{1}{2}ab + \frac{1}{2}ab + \frac{1}{2}c^2$$

$$= ab + \frac{1}{2}c^2.$$

We now equate the two expressions of the area of trapezoid *ABCD* to get:

$$\frac{1}{2}a^2 + ab + \frac{1}{2}b^2 = ab + \frac{1}{2}c^2$$

$$\frac{1}{2}a^2 + \frac{1}{2}b^2 = \frac{1}{2}c^2.$$

By multiplying through by 2, the familiar $a^2 + b^2 = c^2$ emerges, which is the *Pythagorean theorem*.

There are over 400 proofs[9] of the Pythagorean theorem available today; many are ingenious, yet some are a bit cumbersome. However, none will ever use trigonometry. Why is this? As stated earlier, there can be no proof of the Pythagorean theorem using trigonometry since trigonometry depends (or is based) on the Pythagorean theorem. Thus, using trigonometry to prove the very theorem on which it depends would be circular reasoning. Have you been motivated to discover a new proof of this most famous theorem?

UNIT 62

The Golden Ratio — and the Golden Rectangle and Triangle

The Golden rectangle, by its very name, would indicate a beautifully shaped rectangle. Research, largely during the 19th century, by various psychologists, has shown that the most beautiful rectangle is the one that enjoys the following relationship between the length and the width: $\frac{\text{length}}{\text{width}} = \frac{\text{length}+\text{width}}{\text{length}}$. This is known as the Golden Ratio, the rectangle that has this dimensional relationship is shown in Figure 4.29, where rectangle *ABCD* is a *Golden Rectangle*.

Now that we have become somewhat familiar with the Golden Ratio, we can begin to appreciate its application in architecture and art. One time-honored application is its display in the works of the great Greek sculptor, Phidias (ca. 490–ca. 430 BCE). His design for the construction of the Parthenon in Athens, Greece (Figure 4.30) as well as the sculptures he made to adorn this structure, such as the famous statue of Zeus, are said to be reflective of this beautiful ratio. As a matter of fact, the Greek letter φ is used by many mathematicians to represent the Golden Ratio as it is the first letter of Phidias' name when

---

[9]A classic source for 370 proofs of the Pythagorean theorem is Elisha S. Loomis' The Pythagorean Proposition (Reston, VA: NCTM, 1968)

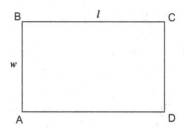

Figure 4.29   Golden rectangle *ABCD*

written in Greek as *φειδίας*.[10] As you can see in Figure 4.30, the Parthenon fits nicely into a Golden Rectangle — that is a rectangle where the quotient of the sides is the Golden Ratio. Furthermore, you will notice several additional Golden Ratios shown in the figure. Yet, even today, no one can say with certainty that Phidias consciously had the Golden Ratio in mind when he designed the structure.

An early significant sighting of the Golden Ratio is found in a three-volume book, entitled *De Divina Proportione* (*The Divine Proportion*), written in 1509 by the Franciscan friar and mathematician, Fra Luca Pacioli (ca. 1445–1514). The book contains drawings of the 5 Platonic solids (see page 195) by the Italian painter, sculptor, architect, and also mathematician, Leonardo da Vinci (1452–1519). The frontispiece is the Vitruvian Man (Figure 4.31), which was drawn in about 1487. This is a picture of a man's body, which clearly exhibits a very close approximation to the Golden Ratio.

Da Vinci provided notes based on the work of Vitruvius (ca. 84 BCE–ca. 27).[11] The drawing, which is in the possession of the Gallerie dell'Accademia in Venice, Italy, is often considered one of the early breakthroughs of pictorially depicting a perfectly proportioned human body.

It appears that da Vinci derived these geometric proportions from Vitruvius's treatise *De Architectura*, Book III.

---

[10] Some mathematics books use the Greek letter τ (tau) representing the first letter of the Greek word "to cut."

[11] **Marcus Vitruvius Pollio** was an ancient Roman writer, architect and engineer.

Figure 4.30   The Parthenon in Athens, Greece

The drawing shows a male figure in two superimposed positions with his arms and legs apart. In one position, with his arms horizontal, he is inscribed in a square, and in the other position, with his arms pointing upward, he is inscribed in a circle. and the two figures are tangent at only one point. The Golden Ratio is exhibited as follows: The distance from the soles of the man's feet to his navel (which appears to be at the center of the circle, as shown in Figure 4.32) divided by the distance from the navel to the top of his head, which is about 1.656, and is close to the Golden Ratio (which we know is 1.618...).

Had the square's upper vertices been somewhat closer to the circle, then the Golden Ratio would have been attained. This can be seen in Figure 4.33, where the radius of the circle is selected to be 1 and the side of the square is 1.618, approximately equal to φ, the Golden Ratio.

Let's examine the algebraic relationship between the length and the width of the Golden Rectangle and show how the ratio is arrived at.

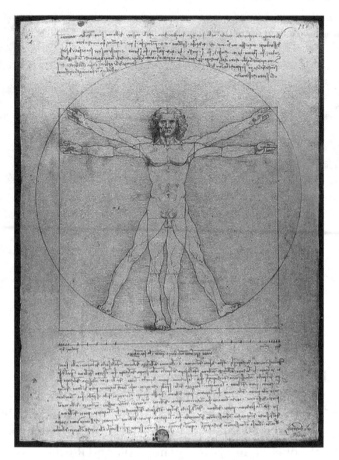

Figure 4.31   Leonardo's Vitruvian man

$$\frac{l}{w} = \frac{l+w}{l}$$

$$\frac{l}{w} = \frac{l}{l} + \frac{w}{l} = 1 + \frac{w}{l} \quad \text{(Separating fractions)}$$

$$\frac{l}{w} = 1 + \frac{w}{1}$$

When we multiply both sides by w, we get $w^2 + w - 1 = 0$. Solving this equation for w by applying the quadratic formula, we find that the positive value of $w = \frac{1+\sqrt{5}}{2} \approx 1.618$. This value is designated by ϕ (phi) and is the Golden Ratio which is the ratio of the length to the width of the Golden Rectangle.

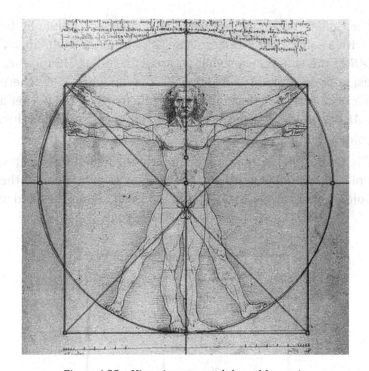

Figure 4.32 Vitruvian man and the golden ratio

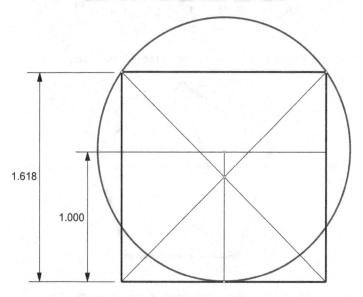

Figure 4.33

The Golden Rectangle can be constructed very simply by beginning with a square $ABCD$ (Figure 4.34) and from the midpoint, M of a side AB of the square, we swing an arc with radius MC to reach the extension of AB at point E, which we show in Figure 4.34, where the Golden Rectangle is then completed by constructing the perpendicular to AE at point E to meet the extension of DC at point F. We then get the Golden Ratio, $\frac{FE}{EB} = \frac{1}{\frac{\sqrt{5}-1}{2}} = \frac{2}{\sqrt{5}-1} = \frac{\sqrt{5}+1}{2}$.

The Golden Ratio appears frequently throughout geometry, for example, in a regular pentagon the diagonals divide each other in the Golden Ratio, as shown in Figure 4.35. This can be verified using

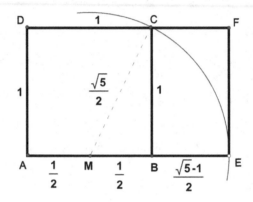

Figure 4.34   Construction of the golden rectangle

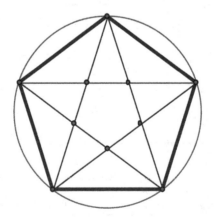

Figure 4.35   Regular pentagon showing golden ratio

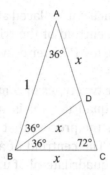

Figure 4.36   Golden triangle

the isosceles triangles formed by the diagonals and their properties.

Consider the isosceles triangle *ABC* in Figure 4.36, whose vertex angle has measure 36°. Then draw the bisector *BD* of ∠*ABC*.

Therefore, ∠*DBC* = 36°, and Δ*ABC* ~ Δ*BCD*. Let *AD* = *x* and *AB* = 1. Since Δ*ADB* and Δ*DBC* are isosceles, *BC* = *BD* = *AD* = *x*. From the similarity above: $\frac{DC}{BC} = \frac{AD}{AB}$, or $\frac{1-x}{x} = \frac{x}{1}$.

This gives us: $x^2 + x - 1 = 0$, and $x = \frac{\sqrt{5}-1}{2}$ (The negative root cannot be used for the length of *AD*) We recall that $\frac{\sqrt{5}-1}{2} = \frac{1}{\phi}$. The ratio for Δ*ABC* of $\frac{\text{side}}{\text{base}} = \frac{1}{x} = \phi$. We therefore call this a *Golden Triangle*.

There are many books written about the Golden Ratio, which manifests itself throughout many aspects of geometry, sometimes in a rather surprising way. We offer two sources here. *The Glorious Golden Ratio*, and *The (Fabulous) Fibonacci Numbers* by A. S. Posamentier and I. Lehmann, Amherst, NY: Prometheus Book, 2012 and 2007, respectively.

UNIT 63

The Center of Gravity of Triangles and Quadrilaterals

You may recall from high school geometry that the center of gravity of a triangle is located at the intersection point of the three medians of the triangle. That means a physical triangle of uniform density will balance on the end of a pencil if it is placed at this median intersection point, also known as the centroid of the triangle. In Figure 4.37, we show the centroid of triangle *ABC* after drawing the three medians to get the point *G*.

We shall now consider two types of centers of a quadrilateral. To find the *centroid* of the quadrilateral is a bit more difficult as it involves repeating 4 times the process that we have used for finding the centroid of a triangle. The centroid of a quadrilateral is also the point on which a physical quadrilateral of uniform density — such as a cardboard quadrilateral — will balance. This point may be found in the following way. In Figure 4.38, let *M* and *N* be the centroids of △*ABC*

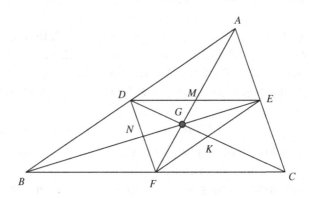

Figure 4.37

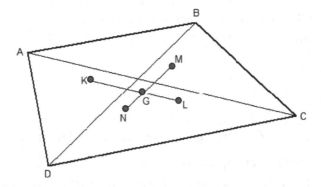

Figure 4.38

and △*ADC*, respectively. Let *K* and *L* be the centroids of △*ABD* and △*BCD*, respectively. The point of intersection, *G*, of *MN* and *KL* is the centroid of the quadrilateral *ABCD*.

The *centerpoint* of quadrilateral *ABCD,* which is the point equidistant from the opposite sides, is the point of intersection, *G*, of the two segments shown in Figure 4.39, *FE* and *HJ*, each joining the midpoints of the opposite sides of the quadrilateral.

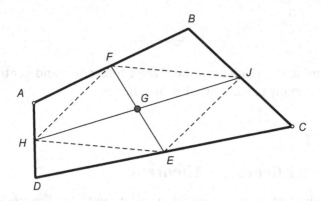

Figure 4.39

We can also see in Figure 4.40 that the lines joining the midpoints consecutively of any random quadrilateral create a parallelogram, which in Figure 4.39 is *FJEH*. This leads us to the fact that the 2 lines joining the midpoints, namely *FE* and *HJ* bisect each other, since the diagonals of a parallelogram are known to bisect each other. The centerpoint also has another unexpected feature. It bisects the segment joining the midpoints of the diagonals of the original quadrilateral. In Figure 4.40, the segment joining the midpoints of diagonals *BD* and *AC*, namely points *M* and *N*, is bisected by the centerpoint *G*.

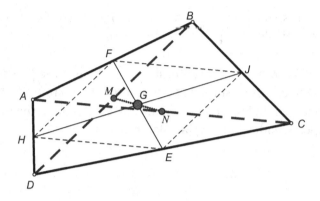

Figure 4.40

An ambitious reader can take the centerpoint and centroid and investigate many of their other properties.

UNIT 64

Napoleon's Geometry Theorem

Perhaps the last thing we would associate with the French Emperor Napoleon Bonaparte (1796–1821) is a geometry theorem that he discovered. To this day there is still some doubt as to whether he was responsible for the discovery of this wonderful geometric relationship, however, it bears his name, nonetheless. Though this relationship can be proved with the barest minimum knowledge of geometry, it is deceptively difficult to prove. How can a theorem be simply expressed and yet difficult to prove? This may sound like a contradiction, but it is not uncommon in mathematics as you will see in this theorem. It's actually rewarding to prove, and the results of the proof of the theorem that is established, is extraordinarily powerful with lots of extensions. In other words, to do the proof can be fun (or at least generate a feeling of accomplishment), but the real nice "stuff" comes once we can work *with* the results.

The preliminary theorem states the following:

The segments joining each vertex of any triangle with the remote vertex of the equilateral triangle drawn externally on the opposite side of the given triangle are equal in length.

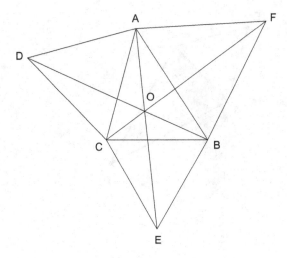

Figure 4.41

That is, in Figure 4.41, where the original triangle *ABC* has the three equilateral triangles drawn on each of its sides, we have *AE* = *BD* = *CF*. Take note of the unusualness of this situation. We start off with *any* triangle and still this relationship holds true regardless of the shape of the triangle. If you were to draw your own triangle you will come up with the same conclusion. Using either a straightedge and a pair of compasses or a dynamic geometry program such as *Geometer's Sketchpad* or *GeoGebra* would clearly demonstrate this relationship.

Before we embark on the adventure that this theorem holds, it may be helpful to give you a hint as to how to prove this theorem. The trick is to identify the proper triangles to prove congruent. They are not easy to identify. One pair of these triangles is shown in Figure 4.42. These two congruent triangles will establish the congruence of *AE* and *DC*. The other segment *BF* can be shown to be congruent in a similar way with an analogous pair of congruent triangles embedded in this figure.

There are quite a few unusual properties in this figure. For example, you may have noticed that the three segments *AE*, *BD* and *CF* are also concurrent where they all intersect in one point. Although the concept of concurrency is not stressed much in the typical high school

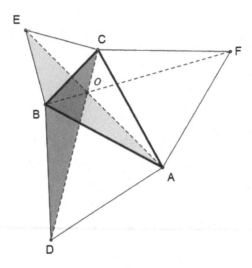

Figure 4.42

geometry course, it should not be taken for granted. It must be proved but for our purposes we shall accept it without proof.[12]

Not only is point O (Figure 4.42) a common point for the 3 segments, but it is also the only point in the triangle where the sum of the distances to the vertices of the original triangle is a minimum. This is often called the *minimum-distance point* of the triangle *ABC*. That is, the sum of the distances from any other point in the triangle to the 3 vertices is greater than from this minimum-distance point.

As if this weren't enough, this point, O, is the only point in the triangle where the sides subtend equal angles. That is, $\angle AOC = \angle COB = \angle BOA$ (see Figure 4.43).

There is more! Locate the center of the each of the 3 equilateral triangles which is the point equidistant from each of the 3 sides as shown in Figure 4.44. You can do this in a variety of ways: find the point of intersection of the three altitudes, medians or angle

---

[12] For a proof of this theorem and its extensions, see A. S. Posamentier, Advanced Euclidean Geometry: Excursions for Secondary Teachers and Students, (Emeryville, CA: Key College Publ., 2002) Chapter 4.

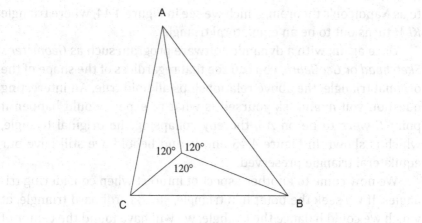

Figure 4.43

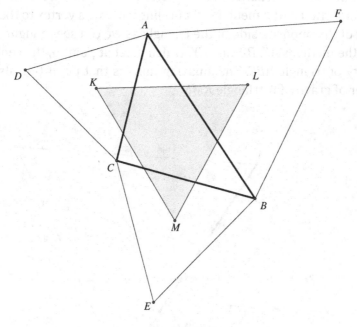

Figure 4.44

bisectors. Joining these center points reveals that an equilateral triangle appears. Remember, we began with just any randomly drawn triangle and now all of these lovely properties appear. This is referred to as Napoleon's theorem, which we see in Figure 4.44, where triangle *KLM* turns out to be an equilateral triangle.

Once again, with a dynamic software program such as *Geometer's Sketchpad* or *GeoGebra,* you can see that regardless of the shape of the original triangle, the above relationships all hold true. An interesting question you might ask yourself is what to expect would happen if point *C* were to be on *AB*, thereby collapsing the original triangle, which is shown in Figure 4.45 and lo and behold, we still have our equilateral triangle preserved.

We now come to another aspect of interest when considering triangles. If we seek the point in a triangle, say, a cardboard triangle, at which we could balance the triangle, we will have found the center of gravity of the triangle, which we have discussed in the previous unit. Recall that for any triangle, the center of gravity is located at the point of intersection of the medians, that is lines joining a vertex to the midpoint of the opposite side of the triangle as we can see in Figure 4.46 with the medians *AX, BZ*, and *CY*, which meet at point *G*, the center of gravity of triangle *ABC*. The amazing thing is that point G is also the center of gravity for triangle *KMN*.

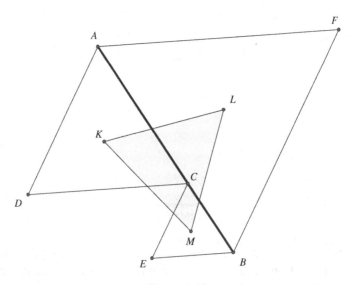

Figure 4.45

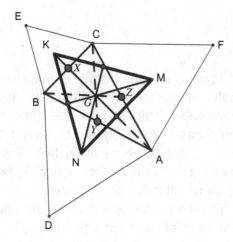

Figure 4.46

Perhaps even more astonishing is the generalization of this theorem. That is, suppose we were to construct similar triangles appropriately placed on the sides of our randomly drawn triangle, and joined their centers (this time we must be consistent as to which "centers" we choose to use — centroid, orthocenter, incenter,[13] etc.), the resulting figure will be similar to the three similar triangles.

Once again, using dynamic geometry software, you can see that all the relationships we have shown above about triangles drawn *externally* on the sides of our randomly selected triangle, can be extended to triangles drawn *internally* as well.

So, from this simple theorem curiously attributed to Napoleon came a host of incredible relationships, many of which can be nicely appreciated using dynamic geometry software. One such finding that an ambitious reader may want to verify is that the circumcircles of each of the equilateral triangles which were drawn on the sides of the original triangle meet at a common point.

---

[13] The centroid is the point of intersection of the medians of a triangle. The orthocenter is the point of intersection of the altitudes of a triangle. The incenter is the point of intersection of the angle bisectors of a triangle.

UNIT 65

Simson's Line

One of the great injustices in the history of mathematics involves a theorem originally published by William Wallace in Thomas Leybourn's *Mathematical Repository* (1799–1800) which, through careless misquotes, has been attributed to Robert Simson (1687–1768), a famous English interpreter of Euclid's *Elements* (see page 206). To be consistent with the historic injustice, we shall use the popular reference, and call it *Simson's Theorem*.

We provide this theorem because it should be a part of the introduction to Euclidean geometry but somehow gets neglected. The beauty of this theorem lies in its simplicity. Suppose you draw a triangle with its vertices on a circle (something that is always possible since any three non-collinear points determine a circle) and select a point on the circle that is not at a vertex of the triangle. From that point you draw a perpendicular line to each of the 3 sides. The 3 points where these perpendiculars intersect the sides (points X, Y and Z in Figure 4.47) are always collinear (i.e. they lie on a straight line). The line that these 3 points determine is often called the *Simson Line*.

We can more formally state Simson's Theorem as follows: *The feet of the perpendiculars drawn from any point on the circumscribed circle of a triangle to the sides of the triangle are collinear.*

In the Figure 4.47, point P is on the circumcircle of $\triangle ABC$. $PY \perp AC$ at Y, $PZ \perp AB$ at Z and $PX \perp BC$ at X. According to Simson's (i.e. Wallace's) Theorem, points X, Y, and Z are collinear thus determining the *Simson Line*.

For the interested ambitious reader, we offer here a quick proof of this collinearity. Since $\angle PYA$ is supplementary to $\angle PZA$ because they are both right angles, quadrilateral $PZAY$ is cyclic.[14] We then draw PA, PB and PC. Therefore $\angle PYZ = \angle PAZ$. Similarly, since $\angle PYC$ is supple-

---

[14] When the opposite angles of a quadrilateral are supplementary (i.e. have a sum of 180°), the quadrilateral is cyclic (i.e. all four points lie on the same circle).

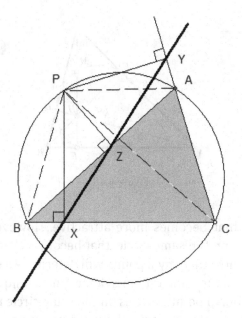

Figure 4.47   Simson line *XZY*

mentary to ∠*PXC*, quadrilateral *PXCY* is cyclic and ∠*PYX* = ∠*PCB*, since they are two inscribed angles intercepting the same arc of the circumscribed circle of quadrilateral *PXCY*. However, quadrilateral *PACB* is also cyclic, since it is inscribed in the given circumcircle and therefore ∠*PAZ* = ∠*PCB*. We can therefore conclude that ∠*PYZ* = ∠*PYX*, and thus points *X*, *Y*, and *Z* are collinear.

This can be nicely demonstrated with dynamic geometry software such as *Geometer's Sketchpad* or *GeoGebra*. There, you would draw the figure and then by moving the point along the circle to various positions, you can observe how the collinearity is preserved under all positions of the point *P*.

UNIT 66

The Nine-Point Circle

In geometry, we are always fascinated when 3 points can be found to lie on the same straight line. When there are more than 3 points on

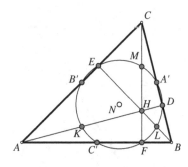

Figure 4.48

the same line, that becomes more attractive. However, when more than 3 points lie on the same circle, that becomes even more appealing. Just as we know that any 2 points will determine a unique line, we know that any 3 noncollinear points determine a unique circle. When we find that a fourth point also lies on the same circle it is quite noteworthy. In this chapter, we will show a fascinating geometric construction that results in nine points lying on the same circle. It is phenomenal! Consider the triangle *ABC* in Figure 4.48. These nine points, for any given triangle, are:

- the midpoints of the sides. In Figure 4.48 they are the points *A′, B′, C′*.
- the feet of the altitudes. In Figure 4.48 they are the points *D, E, F.*
- the midpoints of the segments from the orthocenter (the intersection of the three altitudes of the triangle) to the vertices, which are the points *K, L, M.*

These 9 points all lie on the same circle, which is called the *nine-point circle* of the triangle.

In 1765, Leonhard Euler (1707–1783) showed that six of these points, the midpoints of the sides and the feet of the altitudes, determine a unique circle. It was not until 1820, when a paper[15] published

---

[15] Recherches sur la determination d'une hyperbole équilatèau moyen de quartes conditions données (Paris, 1820).

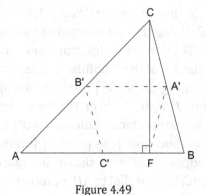

Figure 4.49

by the two French mathematicians Charles Julien Brianchon (1783–1864) and Jean-Victor Poncelet (1788–1867) showed that the remaining 3 points — the midpoints of the segments from the orthocenter to the vertices — were also found to be on this circle. The paper contains the first complete proof of the theorem and initiated the name "the nine-point circle."

To justify that the 9 points lie on the same circle, we shall consider each part with a separate diagram. Bear in mind, though, that each of the following Figures 4.49 through 4.52 is merely an extraction from Figure 4.48, which is the complete diagram. In other words, the following development builds up to Figure 4.52 in small increments.

In Figure 4.49, points A', B', and C' are the midpoints of the 3 sides of $\triangle ABC$ opposite the respective vertex. CF is an altitude of $\triangle ABC$. Since $A'B'$ is a midline of $\triangle ABC$, we have $A'B'||AB$. Therefore, quadrilateral $A'B'C'F$ is a trapezoid. $B'C'$ is also a midline of $\triangle ABC$, so that $B'C' = \frac{1}{2}BC$. Since $A'F$ is the median to the hypotenuse of right $\triangle BCF$, $A'F = \frac{1}{2}BC$. Therefore, $B'C = A'F$ and trapezoid $A'B'C'F$ is isosceles. Recall that when the opposite angles of a quadrilateral are supplementary, as in the case of an isosceles trapezoid, the quadrilateral is cyclic.[16] Therefore, quadrilateral $A'B'C'F$ is cyclic. So far, we have 4 of the 9 points on one circle.

---

[16] A cyclic quadrilateral is one whose four vertices lie on the same circle.

To avoid any confusion, we redraw ΔABC in Figure 4.50 and include altitude AD. Using the same argument as before, we find that quadrilateral A'B'C'D is an isosceles trapezoid and, therefore, it is cyclic. So, we now have 5 of the 9 points on one circle — points A', B', C', F, and D. By repeating the same procedure for altitude BE, we can then state that points D, F, and E lie on the same circle as points A', B', and C'. These 6 points are as far as Euler got with this configuration.

Point H is the orthocenter (the point of intersection of the altitudes), M is the midpoint of CH, as shown in Figure 4.51. Therefore, B'M, a midline of ΔACH, is parallel to AH, or altitude AD. Since B'C' is a midline of ΔABC, we have B'C||BC. We then have right triangle ∠ADC

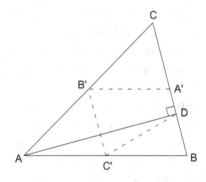

Figure 4.50

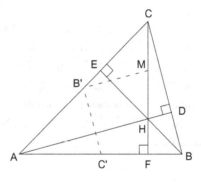

Figure 4.51

so that ∠*MB'C'* is also a right angle. Thus, quadrilateral *MB'C'F* is cyclic (opposite angles are supplementary). This places point *M* on the circle determined by points *B'*, *C'*, and *F*. We now have a seven-point circle.

When we repeat this procedure with point *L*, the midpoint of *BH* as we can see in Figure 4.52, we have as before, ∠*B'A'L* is a right angle, as is ∠*B'EL*. Therefore, points *B'*, *E*, *A'*, and *L* are concyclic (opposite angles supplementary). We now have *L* as an additional point on our circle, making it an eight-point circle.

To locate our final point on the circle, consider point *K*, the midpoint of *AH*. As we did earlier, we find ∠*A'B'K* to be a right angle, as is ∠*A'DK*. Therefore, quadrilateral *A'DKB'* is cyclic and point *K* is on the same circle as points *B'*, *A'*, and *D*. We have, therefore, shown that these *nine specific points* lie on the same circle. This is not to be taken lightly; it is quite spectacular!

In summary: *In any triangle, the midpoints of the sides, the feet of the altitudes, and the midpoints of the segments from the orthocenter to the vertices lie on a circle.*

As with many things in mathematics, there always are additional wonders that can be found when you think you have reached the pinnacle. So as a little extra attraction, we offer what is known as the Euler line, namely, the line joining the orthocenter *H*, the center of the 9-point circle, *N*, and the center of the circumscribed circle, *O*, which we show in Figure 4.53.

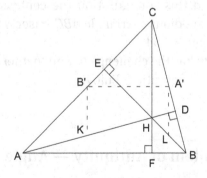

Figure 4.52

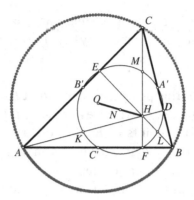

Figure 4.53

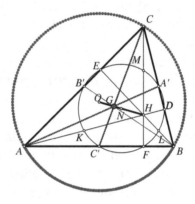

Figure 4.54

Furthermore, in this configuration the centroid, *G,* the point of intersection of the medians of triangle *ABC,* trisects the Euler line *OH,* as we can see in Figure 4.54.

The reader now has the challenge to find further curiosities in this Nine-Point Circle configuration. Enjoy!

UNIT 67

A Famous Problem of Antiquity — Angle Trisection

The 3 famous geometric problems of antiquity which have perplexed mathematicians for centuries still have fame in today's mathematical world. They were posed by the ancient Greeks and remained unsolved for more than two thousand years. And the solution obtained in modern times with the help of algebra was not quite satisfactory either.

These three problems are the following:

(1) *The trisection of an arbitrary angle*

Given a general angle, construct an angle one-third its measure.

(2) *The doubling of a cube*

Given a cube, construct a cube with double the volume.

(3) *The squaring a circle*

Given a circle, construct a square that has the same area as the circle.

The only tools allowed to accomplish these tasks are a pair of compasses and an unmarked straightedge (and a pencil, of course). Moreover, according to Euclidean geometry, the only operations that are allowed to be done with these tools are the following:

(1) Drawing points.

(2) Connecting two points with a line segment.

(3) Drawing a circle centered at a given point with a given line segment as radius.

(4) Marking intersection points (of 2 lines, of a line and a circle, and of 2 circles).

When we say that something can be done with a pair of compasses and straightedge, we mean, in fact, that the whole construction can be reduced to a finite sequence of the steps listed above. It turns out, that the three problems of antiquity are impossible to accomplish with a pair of compasses and straightedge, using only the allowed operations.

The impossibility of these constructions follows from modern algebra and is rather difficult to prove at a high school level. The general result is the following: Starting with a line segment of unit length, a line segment of length L can only be constructed (using a pair of compasses and a straightedge), if L can be obtained from the rational numbers by a finite number of steps involving the operations addition/subtraction, multiplication/division, and taking square roots. Using the theory of fields,[17] one can show that whenever a number L is constructible in this sense, then L is an algebraic number (i.e. a root of a polynomial with integer coefficients), and that the degree of its minimal polynomial is a power of 2 (the minimal polynomial is the polynomial with the smallest degree that has L as a root). Classic problems 1 and 2 (stated above) lead to minimal polynomials of degree 3, while classic problem 3 would need the construction of the square root of π, which is a transcendental number (that is, it is not the root of a non-trivial polynomial equation with integer coefficients).

Thus, all these problems have been proven to be unsolvable. Of course, this result depends on the required exclusive usage of the compasses and straightedge for the construction. It is possible to obtain solutions, if other tools are allowed, or if the tools are used in non-standard ways.

Here we will provide an example for a trisection of an angle, which had been described around 320 CE by Pappus of Alexandria (ca. 290–ca. 350 CE) but is certainly much older (probably the oldest known construction of this type).

We are given an angle with vertex at P, as indicated in Figure 4.55. Draw a point A on one of the legs of the angle, thus creating a line segment PA of length a. Through point A, draw one line g parallel to the other leg of the angle, and one line h perpendicular to g.

Next, draw a line through the vertex P intersecting line h at point C and line g at point B in such a way, that the length of the segment CB is precisely 2a. (This step can be achieved if we mark the length 2a on a ruler and move the ruler in the plane until it has the required position — unfortunately, this is *not* one of the 4 allowed Euclidean operations mentioned above.)

---

[17] In mathematics, a *field* is a set of rational and real numbers on which the four basic arithmetic operations: addition, subtraction, multiplication and division hold true.

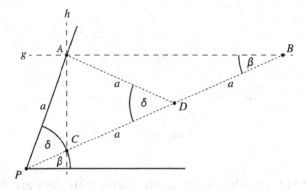

Figure 4.55

The line segment *PB* cuts the given angle into two parts, which we call β (beta) and δ (delta). Obviously, the angle at *B* also equals β.

The midpoint *D* of segment *BC* defines an isosceles triangle *DAP*, so that the angle δ also appears at *D*, as indicated. But the triangle *ADB* is also isosceles, so that the angle δ is easily shown to be 2β. Hence, the original angle at *P* is $\beta + \delta = 3\beta$. This shows that the newly constructed angle β trisects the given angle at *P*.

Another method was developed by Archimedes: Given an angle with vertex *P*, draw a circle with radius *r* around *P*, as shown in Figure 4.56. Let *A* be the intersection of that circle with one of the legs of the given angle. Through *A*, draw a straight line outwards towards BC, such that the segment *CB* has precisely the length *r*. Then the indicated angle at *C* is precisely one third of the given angle at *P*.

Fitting a line segment of a given length between 2 given lines in a certain way is called a "Neusis construction." It requires that 2 points on a straightedge are marked, and the straightedge is moved until the 2 points have the desired position. Unfortunately, this is also not an allowed Euclidean operation.

These famous classic problems have attracted amateurs who tried to find solutions, ignoring the proven fact that they cannot be solved with the classic tools, namely a pair of compasses and an

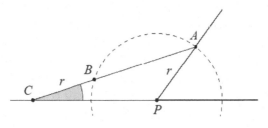

Figure 4.56

unmarked straightedge. Even today, mathematicians are frequently approached by people who believe that they have a solution to one of these famous problems. What they really have is most probably either an approximate or a Neusis construction.

UNIT 68

The Ever-Present Equilateral Triangle

In a high school English class, it is common for a teacher to include the works of the American author Christopher Morley (1890–1957). However, high school mathematics teachers are more apt to mention his father, the mathematician Frank Morley (1860–1937), whose fame today is largely based on a beautiful discovery that he made in 1904, but he didn't publish it until 1924, while he was in Japan. To really appreciate the beauty of this theorem, you should examine it with a dynamic geometry program such as *Geometer's Sketchpad* or *GeoGebra*. We will do the best we can for you to appreciate it here on these pages. The theorem states that:

> *The intersections of the adjacent angle* trisectors[18] *of any triangle meet at three points determining an equilateral triangle.*

---

[18] Here we refer to the two adjacent angle trisectors (the rays that divide an angle into three equal parts) nearest a side of the triangle.

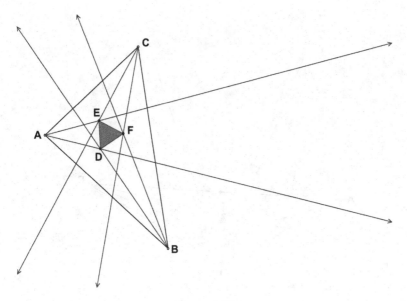

Figure 4.57

Notice in Figure 4.57 how the points *D, E,* and *F* are the intersection points of the adjacent trisectors of the angles of △*ABC*. Then surprisingly, △*DEF* is an equilateral triangle. The amazing aspect of this theorem is that this equilateral triangle will be formed from a triangle of *any shape.* Dynamic geometry programs allow us to change the shape of the original △*ABC* and observe that △*DEF* remains equilateral, although of different size.

We offer a few variations (Figures 4.58–4.60) that will allow you to witness this remarkable relationship. This is truly one of the most surprising relationships in geometry. Be cautioned, the proof is one of the most difficult in Euclidean Geometry. One source for three proofs of this theorem can be found in A. S. Posamentier and I. Lehmann, *The Secrets of Triangles*, Prometheus Books, pp. 351–355.

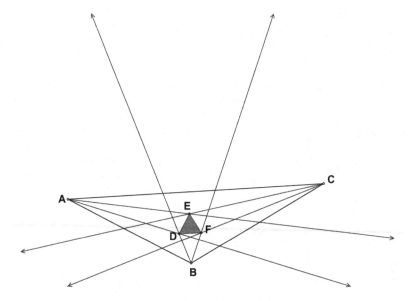

Figure 4.58

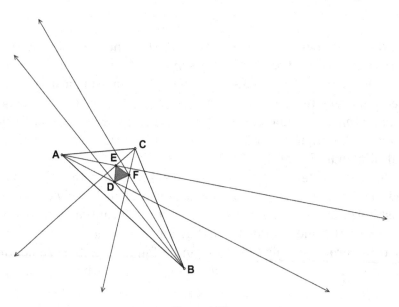

Figure 4.59

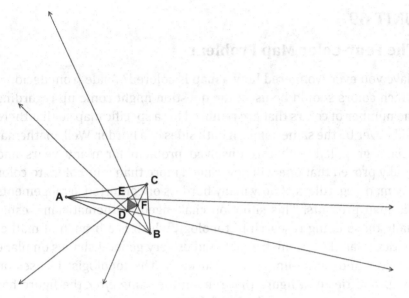

Figure 4.60

Before leaving this highly unexpected relationship, one can add another curiosity which manifests itself in this diagram. If we join each vertex of the original triangle with an opposite vertex of the newly formed equilateral triangle, we find that these 3 lines all intersect in the same point as we can see in Figure 4.61, with the lines *AF*, *BE*, and *CD* being concurrent.

An ambitious reader may seek other curious relationships within this configuration.

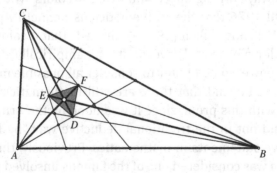

Figure 4.61

UNIT 69

The Four-Color Map Problem

Have you ever wondered how a map is colored? Aside from deciding which colors should be used, the question might come up regarding the number of colors that are required for a specific map so that there will never be the same color on both sides of a border. Well, mathematicians grappled with this unsolved problem for many years and finally proved that one will never need more than four colors to color any map, regardless of how many borders or contorted arrangements the map presents. This situation challenged mathematicians, especially those doing research in topology, which is a branch of mathematics related to geometry that studies very general shapes on plane surfaces and three-dimensional surfaces. The topologist focuses on the properties of a figure that remain the same *after* the figure has been distorted or stretched according to a set of rules. A piece of string with its ends connected may take on the shape of a circle, a square or any polygon, which is all the same for the topologist. In going through this transformation, the order of the "points" along the string does not change. This preservation of ordering survives the distortion of shape, and it is this property that attracts the interest of topologists. Therefore, a circle and all polygons represent the same geometric concept to the topologist.

Throughout the 19th century, it was believed that 5 colors were required to color even the most complicated looking map. However, there was always strong speculation that 4 colors would suffice. It was not until 1976 that the mathematicians Kenneth Appel (1932–2013) and Wolfgang Haken (1928–) "proved" that 4 colors were sufficient to color any map. However, it was empirically done as they used a high-powered computer to consider all possible map arrangements. It must be said that there are still mathematicians who are dissatisfied with this proof, since it was done by the brute force of a computer and not in the traditional deductive way "by hand" using logic to prove statements in mathematics. Previously, this *four-color map problem* was considered one of the famous unsolved problems of

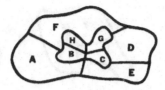

Figure 4.62

mathematics. Let us now consider various maps and the number of colors required to color them in such a way that no common boundary of two regions shares the same color on both of the sides. This is clearly a requirement for coloring any map. Suppose we consider a geographic map that has a configuration analogous to that shown in Figure 4.62.

Here we notice that there are 8 different regions indicated by the letters shown in Figure 4.62. Suppose we list all regions that have a common boundary with region H, and the regions that share a common vertex with region H. The regions designated by the letters B, G, and F share a border with region H. The region designated by the letter C shares a vertex with the region designated by the letter H. Remember, a map will be considered correctly colored when all the regions are completely colored and any 2 regions that share a common boundary have different colors. Two regions sharing a common vertex may also share the same color. Therefore, only 4 colors are needed in this case so all regions bordering region H are different colors.

Let's consider coloring a few maps such as those shown in Figure 4.63 in order to see various configurations a map can have that requires not more than 3 colors (*r*/red; *y*/yellow; *g*/green).

The first map is able to be colored in 2 colors: yellow and red. The second map requires 3 colors: yellow, red and green. The third map has 3 separate regions, but only requires 2 colors, red and green, since the innermost territory does not share a common border with the outermost territory.

One could conclude that if a three-region map can be colored with fewer than 3 colors, a four-region map can be colored with fewer than 4 colors. Let's consider such a map.

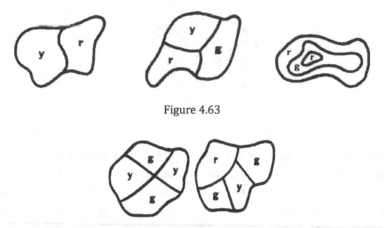

Figure 4.63

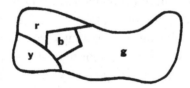

Figure 4.64

The left-side map shown in Figure 4.64 has 4 regions and requires only 2 colors for correct coloring whereas, the right-side map in Figure 4.64 also consists of 4 regions but requires 3 colors for correct coloring.

We will now consider a map that requires 4 colors for proper coloring of the regions. Essentially, this will be a map, where each of the 4 regions shares a common border with the other 3 regions. One possible such mapping is shown in Figure 4.65.

Figure 4.65

If we now take the next logical step in this series of map-coloring challenges, we should come up with the idea of coloring maps involving 5 distinct regions. It will be possible to draw maps that have 5 regions, which require 2, 3, or 4 colors but not more. The task of drawing a five-region map that *requires* 5 colors for correct coloring is not possible. This curiosity can be generalized through further investigation and should help to convince you that any map on a plane surface, with any number of regions, can be successfully colored with 4 or fewer colors.

This challenge remained alive for many years, challenging some of the most brilliant minds, but as we said earlier, the issue has been closed by the work of the two mathematicians, Appel and Haken. There are still many conjectures in mathematics that have escaped proof but have never been disproven. Here, at least, we have a conjecture that has been proved, albeit using modern technology.

UNIT 70

Optical Illusions

We can make mistakes with our visual assessment of a geometric figure. We present some of these optical tricks, as they can be useful to make a person more discriminating with visual presentations. First, we will show some of these easily mistaken assessments, and then show how logical mistakes can be compounded and overlooked. So, follow along as we explore some of these counterintuitive characteristics, which can lead to some geometric mistakes!

Optical Mistakes

We can begin by comparing the two segments in Figure 4.66. The one on the right side looks longer. In Figure 4.67, the bottom segment also looks longer. In actuality, the segments have the same length.

In Figure 4.68, the crosshatched segment appears longer than the clear one, and in Figure 4.69, the right-side figure, the narrower

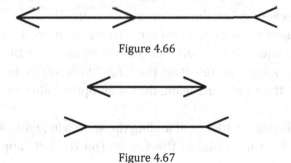

Figure 4.66

Figure 4.67

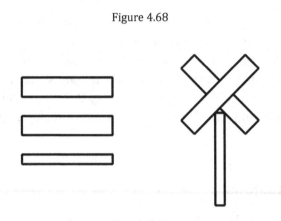

Figure 4.68

Figure 4.69

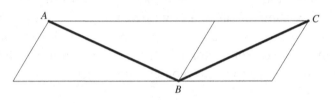

Figure 4.70

vertical stick appears to be longer than the other two crossed sticks, even though to the left they are shown to be the same length.

A further optical illusion can be seen in Figure 4.70, where *AB* appears to be longer than *BC*. This is not true, since *AB* = *BC*.

In Figure 4.71, the vertical segment clearly appears longer but isn't. The curve lengths and curvature of the diagrams in Figure 4.72 are quite deceiving. Yet, the curves are congruent!

The square between the two semicircles in Figure 4.73 looks bigger than the square to the left, but the two squares are the same size. In Figure 4.74, the square within the large black square looks smaller than that to the right, but, again, that is an optical illusion, since they are the same size.

We see further evidence of fooling the senses in Figure 4.75 where the larger circle inscribed in the square (on the left) appears to be

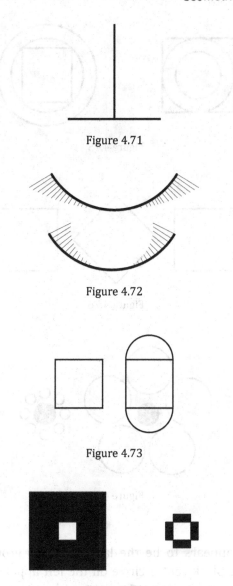

Figure 4.71

Figure 4.72

Figure 4.73

Figure 4.74

smaller than the circle circumscribed about the square (on the right). Again, the circles are the same size!

Figures 4.76, 4.77, and 4.78 show how relative placement can affect the appearance of a geometric diagram. In Figure 4.76, the

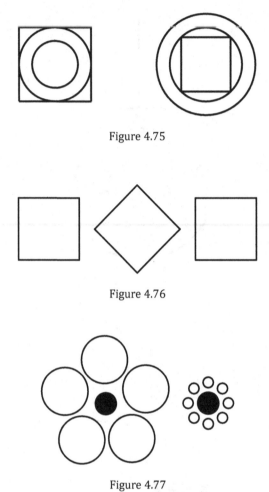

Figure 4.75

Figure 4.76

Figure 4.77

center square appears to be the largest of the group but isn't. In Figure 4.77, the black center circle on the left appears to be smaller than the black center circle on the right, and again it is not.

In Figure 4.78, the center sector on the left appears to be smaller than the center sector on the right. In all these cases, the two figures that appear not to be the same size are, in fact, the same size!

Our journey through these optical illusions should make the reader more alert when looking at geometric figures.

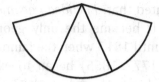
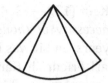

Figure 4.78

UNIT 71

The World of Non-Euclidean Geometries

The geometry that is taught in school is two-dimensional plane geom-etry, called Euclidean geometry, and named after the ancient Greek mathematician Euclid (ca. 325 BCE) who was born in the mid-4th cen-tury BCE and lived until the mid-3rd century BCE. Euclidean geometry only applies to a flat surface, but though the world is spherical, it is still very useful, because we live on such a small piece of it, that its curvature is almost nonexistent, and it can be treated as a flat surface.

In about 300 BCE Euclid wrote the *Elements,* a book which became one of the most famous books ever written. Euclid stated five postu-lates on which he based all his theorems:

1. *To draw a straight line from any point to any other.*

2. *To produce a finite straight line continuously in a straight line.*

3. *To describe a circle with any center and distance.*

4. *That all right angles are equal to each other.*

5. *That, if a straight line falling on two straight lines make the interior angles on the same side less than two right angles, if produced indefinitely, meet on that side on which are the angles less than the two right angles.*

Euclidean geometry became almost a religion in mathematics and it was heretical not to accept it. The influential philosopher

Immanuel Kant (1724–1804) stated that: *Euclidean geometry is the inevitable necessity of thought.* It became the only geometry for over 2000 years, and it was not until 1817 when the famous mathematician Carl Friedrich Gauss (1777–1855) began to experiment with other geometries, which did not depend on the famous fifth postulate stated above. The fifth postulate is essential to Euclidean geometry, and is equivalent to what is known as Playfair's Axiom which can replace the fifth postulate, since it is much easier to understand:

> *Given a straight line and a point not on the line, it is possible to draw exactly one line through the given point parallel to the given line.*

Gauss discovered two other non-Euclidean geometries when he assumed that either more than one, or no lines can be drawn parallel through a given point not on a line. However, being a well-known mathematician, and not wanting to cause controversy, he did not publish his work.

Curiously the young brilliant Hungarian mathematician Janos Bolyai (1802–1860) wrote Gauss that he had discovered these non-Euclidean geometries and Gauss urged him to publish these concepts which he independently developed, relieving Gauss of any embarrassment.

Let us first consider the geometry where no parallel lines can be drawn through a point not on a line. A prime example of this type of geometry is spherical geometry, which is discussed on page 264. Before we can show on a sphere how no parallel lines can be drawn through a point not on a line, we have to explain which lines on a sphere are analogous to straight lines on a plane.

What defines a straight line on a plane is that it is the shortest distance between two points. Therefore, a "straight line" on a sphere, called a *geodesic*, is a line that is the shortest distance between two points on the sphere. A geodesic on a sphere is the arc of a great circle, which is a circle that divides the sphere exactly in half. The equator on the earth is a great circle, as are the lines of longitude that pass through the North and South Poles. On a sphere, to show that no parallel lines can be drawn through a point not on a line, is equivalent to show that no parallel great circles can be drawn through a point not on a great circle. To illustrate this, consider the great circle of the equator and the North pole, which is a point not on the equator. Every great circle that can be drawn through the North Pole is a circle of longitude which intersects the equator and, therefore, cannot be parallel to it (see Figure 4.79).

The sphere is a convex surface and is said to have constant positive curvature. Another example of a surface whose geometry is similar to a sphere is an ellipsoid, which is the surface generated by an ellipse rotated around its major axis and can look similar to a football. See Figure 4.80 which shows the geodesics on the ellipsoid, which are the darkened lines.

Not only are there no parallel lines on a positively curved surface but triangles and quadrilaterals on a convex surface have very different properties than ones on a plane surface. As shown in the next unit

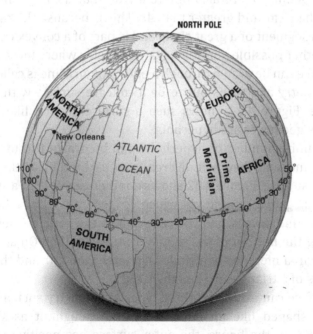

Figure 4.79   The equator and lines of longitude

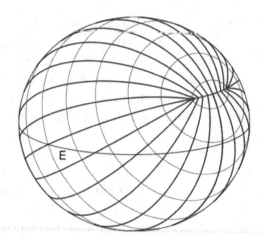

Figure 4.80   Ellipsoid

on spherical geometry (page 000), the sum of the angles of a spherical triangle is between 180° and 540°. The circumference of circles in spherical geometry are also different from Euclidean geometry and are less than $2\pi r$ and greater than $4r$. This is because the radius of a circle is a segment of a great circle and is part of a convex curve.

The other possible non-Euclidean geometry where more than one parallel line can be drawn through a point not on a line is called *hyperbolic geometry* and turns out to be a concave surface with negative curvature. Figure 4.81 shows such a surface shaped like a saddle where the geodesics are hyperbola.

An infinite number of hyperbolas can be drawn through a point not on a line, and they will diverge from each other as they extend along the surface. Figure 4.82 shows another example of a surface of constant negative curvature, which is generated by a hyperbola rotated around its axis, and you can see how the parallel lines diverge traversing the bottom of the saddle from each other. On such a surface, the circumference of a circle is greater than $2\pi r$ and the sum of the angles of a triangle are less than 180°.

A surface can exhibit all three types of geometry such as a Torus which is shaped like an inner tube or a doughnut as shown in Figure 4.83. In the figure, the outer surface has positive curvature

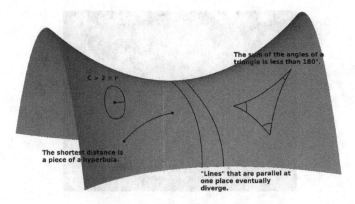

The sum of the angles of a triangle is less than 180°.

C > 2 π r

The shortest distance is a piece of a hyperbola.

"Lines" that are parallel at one place eventually diverge.

Figure 4.81   Hyperbolic geometry

Figure 4.82   Surface of constant negative curvature

while the inner surface has negative curvature, and the very top of the Torus where the curvature changes, has zero curvature like a plane.

Bolyai's work on non-Euclidean geometry naturally took many years to be fully accepted and is still not commonly taught in school where Euclidean geometry still reigns. However, Einsteinian theory predicts that the universe could be positively or negatively curved, but it is probably mostly flat, which continues to be a source of debate.

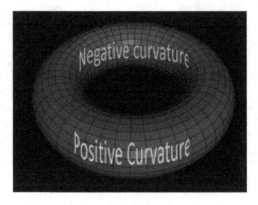

Figure 4.83   Torus

UNIT 72

Introduction to Spherical Geometry

When we speak of geometry, we typically think of the geometric figures that we can draw on a flat piece of paper. This two-dimensional geometry has guided us very nicely through our school mathematics. It has given us a good feel for geometric principles and perhaps also some of the wonders embedded in two-dimensional geometry. However, we live in a world with three spatial dimensions — length, width and depth. It takes three numbers to describe one's exact position at any given moment. All too often we take our three-dimensional world for granted because we live on a very small part of it which is practically flat, and do not consider how it can be described and explained geometrically. For example, when sitting in an airplane crossing the Atlantic, say from New York to Vienna, Austria, the flight pattern shown on a video screen has the plane traveling on what appears to be a curved path crossing relatively near the tip of Greenland. One would think that the pilot is trying to avoid something because he does not appear to be traveling in a "straight line" as envisioned on the video screen. Yet in fact, the pilot is actually choosing the shortest path. The geometry on a sphere, in this case the earth's surface, requires us to delve into spherical geometry, which very easily explains this and many other aspects of our three-dimensional world — including movement above the earth such as satellites and our GPS monitors. Looking at a conventional (plane) map makes us think the southern coast of Florida is the closest to Africa of any point on the East Coast of the US, however, the tip of Maine is actually closer to the coast of Africa. As we begin our journey through three-dimensional geometry, we will provide many aspects of our spherical world that will be easily understood with merely a basic high school mathematics background.

One can classify the circles drawn on the surface of a sphere into two types. There are circles that have their center at the center of the sphere, and other circles whose center does not coincide with the center of the sphere. We will concern ourselves only with the former, which are known as *great circles*, as they are the largest circles that can be drawn on the sphere. Although there are many similarities between the geometry that can be done on a sphere and the geometry that is typically done on a plane, there are also some significant differences that can challenge our thoughts because of their counterintuitive concepts.

The geometry on the surface of the sphere is known as *spherical geometry*, while the geometry on a plane is called *plane geometry*. One notable difference is that there are no parallel lines in spherical geometry.

As we begin to investigate the geometry of the sphere, we need to set up some ground rules. For example, on a plane, we know that the shortest distance between two points is a straight-line segment joining them, while on the sphere it turns out that the shortest distance between two points is the arc of a great circle joining them. Hence these are considered the "straight lines" on a sphere and are also called *geodesics*. Any other line on a sphere that is not the arc of a great circle would be analogous to a curved line on a plane. As we mentioned earlier, it is sometimes puzzling why an airplane flight from New York to Vienna, Austria, typically flies north near Greenland. When looking at a map that is plane projection of the earth, a shorter-appearing distance would be a "straight line" over the Mid-Atlantic

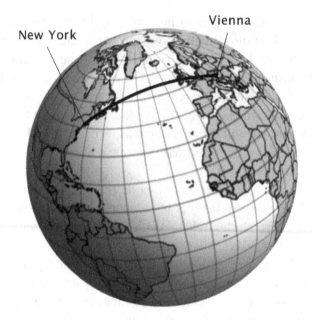

Figure 4.84   Great circle route from New York City to Vienna

Ocean. In fact, the shorter distance — the great circle route — is the way the airplanes typically fly, closer to Greenland (Figure 4.84).

On the Earth's surface, there are two prominent types of great circles: the equator and the longitudinal lines. Great circles on the same sphere have a number of noteworthy properties. Some of these are listed here.

- The axis of a great circle (which is the line through the center of the circle and perpendicular to the plane of the sphere) on the sphere passes through the center of the sphere.
- All great circles of the same sphere are equal.
- Every great circle bisects a sphere.
- Any two great circles of the same sphere bisect each other.
- If the planes of two great circles of the same sphere are perpendicular to each other, each of the circles passes through the poles of the other. (The poles of a great circle are where the line drawn through the center and perpendicular to the plane of the great

circle intersects the sphere — as the North Pole and the South Pole are to the great circle of the equator.)

- Through any two points on a sphere (excluding the endpoints of a diameter) one and only one great circle can be drawn.

We are now ready to consider a spherical angle. When the arcs of two great circles meet on the sphere, they determine a spherical angle, which is measured by the angle formed by the tangents drawn to each great circle at the vertex of the spherical angle. We show this in Figure 4.85 as $\angle A'PB'$, whose vertex is P and has the same measure as the arc of the great circle AB

We also notice in Figure 4.85, that the angles formed by the arcs PA and PB are drawn to form right angles on the sphere at points A and B. In general, we would say that all arcs of great circles drawn through the pole of another given great circle form right angles with the given circle.

Note that the words we use in plane geometry referring to straight lines, such as *diagonal, altitude, median, bisector* have the same relationship to spherical polygons as straight lines have to polygons

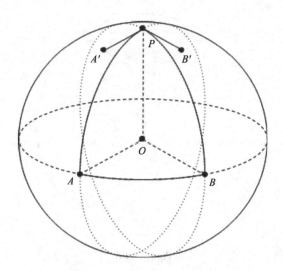

Figure 4.85 Spherical angle

drawn on a plane. Similarly, the words *right, obtuse, acute, equilateral, isosceles*, and *equiangular* also apply to spherical triangles. Remember, the sides of spherical triangles are all arcs of great circles on the sphere, since those are analogous to the straight lines in plane geometry.

Analogous to the triangle in the plane, the sum of any two sides of a spherical triangle is greater than the third side of the spherical triangle. In Figure 4.86, we can conclude that for the sides of spherical triangle *ABC, AB + BC > AC, AC + BC > AB*, and *AB + AC > BC*.

Furthermore, the sum of the sides of any convex spherical polygon is less than 360°, since if the sum of the sides were equal to 360°, then it would be a great circle. In Figure 4.87, we can see that the sum of the sides of the spherical quadrilateral, *AB + BC + CD + DA* < 360°, which are measured by the angles at the center of the sphere. Clearly, you can see that if the sum were equal to 360°, a great circle would be formed.

Perhaps one of the places where a spherical triangle departs most radically from a plane triangle is that the sum of the angles of a spherical triangle must be greater than 180° and less than 540°.

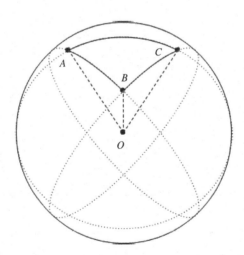

Figure 4.86

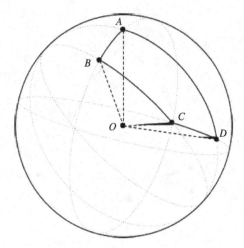

Figure 4.87

Another way of looking at it, would be that a spherical triangle can contain one, two or three right angles, as well as it can contain one, two, or three obtuse angles. If a spherical triangle had three angles of 180° then it would be a great circle and the triangle would disappear. Of course, none of this would be true with a plane triangle, whose angle sum is always 180°.

For us to demonstrate the angle sum of a spherical triangle, we first need to define a polar triangle. If the vertices of a spherical triangle are the poles of the sides of a second spherical triangle, then the two triangles are *polar triangles* of each other. In Figure 4.88, the points A, B, and C are each the poles of arcs $B'C'$, $A'C'$ and $A'B'$ respectively. Likewise, the points A', B' and C' are the poles of the sides of spherical triangle ABC, specifically, BC, AC, and AB, respectively.

We would also need to establish another relationship between two polar triangles. That is, for any two polar triangles any angle of one is supplementary to the opposite side of the other polar triangle. In other words, in Figure 4.89, we have $\triangle A'B'C'$ is the polar triangle of $\triangle ABC$. We will show that

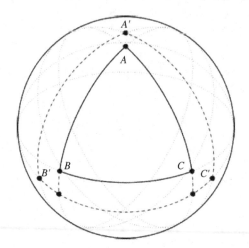

Figure 4.88

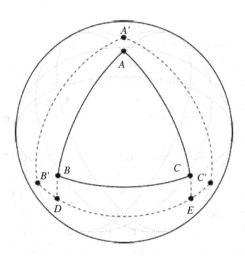

Figure 4.89

$\angle A + B'C' = 180°$, and $\angle A' + BC = 180°$

$\angle B + C'A' = 180°$, and $\angle B' + CA = 180°$

$\angle C + A'B' = 180°$, and $\angle C' + AB = 180°$.

We begin our demonstration by extending arcs AB and AC to meet $B'C'$ at points D and E respectively. Note, that B' is the pole of AE, and $B'E = 90°$. Similarly, $DC = 90°$. Therefore, $B'E + DC' = 180°$. We can see that $B'E = DE + B'D$, which allows us to conclude that $DE + B'D + DC' = 180°$. Since A is the pole of DE, we know that $DE = \angle A$. We can then conclude that $\angle A + B'C' = 180°$. The same argument can be made for each of the other angles the spherical triangle, showing the supplementary relationship between an angle of one spherical triangle and the opposite side of its polar triangle.

We are now ready to demonstrate how the sum of the angles of a spherical triangle is greater than 180° and less than 540°. To begin, we are given spherical triangle ABC and its polar triangle $A'B'C'$, where, as shown in the Figure 4.90, the number of degrees of each of the sides of this polar triangle are indicated as a', b' and c'.

Since earlier, we established the supplementary relationship between an angle of one spherical triangle to the opposite side of its polar triangle, we can conclude the following:

$\angle A + a' = 180°$, $\angle B + b' = 180°$, $\angle C + c' = 180°$. If we now take the sum of these three equations, we get: $\angle A + \angle B + \angle C + a' + b' + c' = 540°$. We know that the sum of three sides of a spherical triangle is less than 360°; thus, $a' + b' + c' < 360°$. Therefore, subtracting this from the

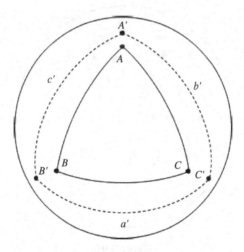

Figure 4.90

previous sum, we can conclude that the sum of the angles of a spherical triangle must be greater than 180°, or for our triangle ABC, we can conclude that $\angle A + \angle B + \angle C > 180°$.

Once again using the relationship that $\angle A + \angle B + \angle C + a' + b' + c' = 540°$, and it surely makes sense that $a' + b' + c' > 0°$, we can conclude that $\angle A + \angle B + \angle C < 540°$. Therefore, we have justified our original statement that $180° < \angle A + \angle B + \angle C < 540°$.

Spherical triangles can also be congruent. However, in comparison to plane triangles, spherical triangles can have corresponding parts equal and not be congruent because of their orientation. In this case, they are referred to as *symmetric triangles*. We can see congruent spherical triangles (Figure 4.91) and symmetric spherical triangles (Figure 4.92).

As with plane triangles, there are congruence relationships such as the following:

- Two spherical triangles on the same sphere, or on equal spheres, are congruent (or symmetric), if two sides and included angle of one are equal, respectively, to two sides and the included angle of

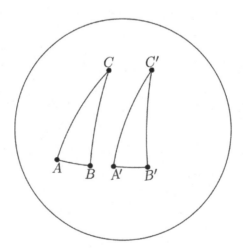

Figure 4.91

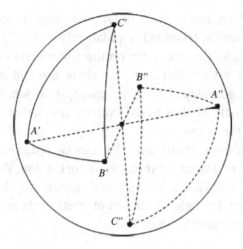

Figure 4.92

the other and arranged in the same order (or opposite order, for the symmetric relationship).

- Two spherical triangles on the same sphere, or on equal spheres, are congruent (or symmetric), if two angles and included side of one are equal, respectively, to two angles and included side of the other and arranged in the same order (or opposite order, for the symmetric relationship).
- Two spherical triangles on the same sphere, or on equal spheres, are either congruent or symmetric if the three sides of one are equal, respectively, to the three sides of the other.
- Two spherical triangles on the same sphere or on equal spheres are either congruent or symmetric, if the three angles of one are equal, respectively to the three angles of the other.

We also know that two symmetric spherical triangles — and obviously, congruent spherical triangles — are equal in area. Furthermore, analogous relationships that exist in a plane isosceles triangle, also exist for a spherical isosceles triangle. That is, the base angles of a spherical isosceles triangle are equal, and conversely if the base angles of a spherical triangle are equal, then the sides opposite them are also equal. Also, a spherical equilateral triangle is equiangular, and equiangular spherical triangle is equilateral.

As you can see, the geometry on the sphere is a complete study, where there are no parallel lines, but there are still many comparisons to plane geometry. This is an example of what we call non-Euclidean geometry that is based entirely on circles — in this case great circles of the sphere.

There are some conundrums that can tie spherical geometry to plane geometry in a somewhat entertaining fashion. We can see once again in Figure 4.93 that a triangle on a sphere, which we have been calling a spherical triangle composed of great circle arcs as sides, will have an angle sum greater than 180°.

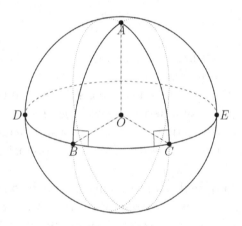

Figure 4.93

By contrasting spherical geometry with plane geometry, we get a first view of what is called a *non-Euclidean geometry* (see the previous unit). This is a geometry where the parallel postulate — namely that, given a point, one and only one line can be drawn through this point and parallel to a given line — does not hold. In other words, the concept of parallelism does not hold on the sphere. One could argue that this geometry would be the more appropriate geometry for us to use since we live on a sphere. As we mentioned earlier, one place where we would notice this spherical phenomenon is in the flight path of and also the sail path of an ocean liner. Recall the example we mentioned earlier, where the flight of an airplane traveling from New York to Vienna, Austria, will travel closer to the southern tip of Greenland, and not as one would expect a "straight line" on a plane map.

UNIT 73

Cartography is Mathematically Challenging

We have all seen the common flat map of the world where the North and South poles are not shown and land near the poles appears very large. Mathematicians struggled for centuries to solve the problem of mapping points from a curved surface to a plane, so that distances, directions, and areas were preserved, but were not successful. The famous German mathematician Carl Friedrich Gauss (1777–1855) solved the problem in 1827 with his *Theorema Egregium* (remarkable theorem) which proved it impossible to construct a plane map of a spherical surface so that the distances, directions, and areas are all true. As shown in the unit on non-Euclidean geometry (page 259), a sphere possesses a very different geometry than a plane. Any flat map of our spherical world will always contain distortions and a cartographer has to decide which property is important to be preserved for the intended purpose.

During the 15th century, the world was being discovered by the great Portuguese navigators who sailed around Africa and to the Artic, culminating in Columbus' voyage to the Americas in 1492 and resulting in maps of the known world that were being continually updated. The man who is most credited for the development of modern cartography is the Flemish mapmaker, Geradus Mercator (1512 –1594) who produced many very accurate local maps. In 1569, his most famous world map was constructed, containing a unique grid design, and became known as the *Mercator projection*. Mercator was able to design the map so that compass directions could be drawn as straight lines which was an issue that other mapmakers struggled with. He accomplished this by first projecting the

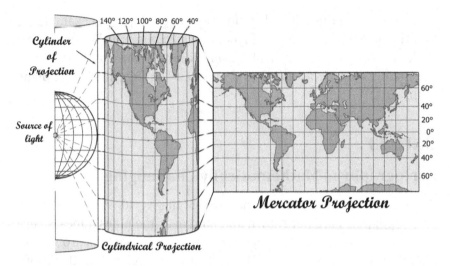

Figure 4.94

points of the sphere onto a cylinder as shown in Figure 4.94. The sphere is actually inside the cylinder and rays are projected from the center of the sphere mapping points from the sphere to the cylinder. The cylinder is then unwrapped into a flat plane producing the Mercator projection. The areas and distances within 15° of the equator are almost true but as one moves further away from the equator the distortion increases, the circles of latitude become further and further apart, and the poles are projected to infinity so cannot be shown at all.

What is preserved, however is the compass bearing of the route between any 2 points, which is called the *Rhumb line*, and is the course sailors generally follow over short distances.

In the 16th century, mathematics had not progressed enough to precisely formulate the cylindrical projection which requires advanced calculus to accurately locate the coordinates of the points that are projected from the cylinder to the plane. Calculus was developed in the 17th century, but it took another century before the three-dimensional geometry of map projection could be mathematically defined. As a result, Mercator had to apply many small steps to a

Figure 4.95 Early Mercator Projection of the Americas

continuous process to achieve his result. Figure 4.95 shows an early Mercator projection of the Americas and Figure 4.96 shows the 1569 Mercator projection of the world.

There are a large number of map projections,[19] each designed for a different purpose, but the most common today is the Mercator projection because it depicts the world where most people live without too much distortion. In particular, if the projection is used to encompass a small part of the Earth, then the result is a very accurate map.

---

[19] See https://en.wikipedia.org/wiki/List_of_map_projections.

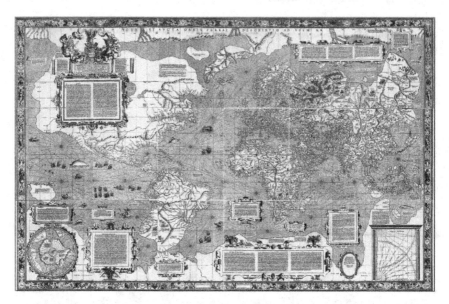

Figure 4.96   Mercator 1569 world map (Nova et Aucta Orbis Terrae Descriptio ad Usum Navigantium Emendate Accommodata)

UNIT 74

A Look at Fractals

The term *fractal* entered into the lexicon recently, aided by computer imaging in 1975. It is derived from the Latin word *fractus,* which can be defined as broken or shattered glass. The Polish born French-American mathematician Benoit Mandelbrot (1924–2010) coined the term to describe shapes of infinite irregularity whose geometric pattern looks similar as you zoom in to smaller and smaller scales. In nature, examples of such shapes are approximated by edges of snowflakes, ocean waves, coastlines and leaf systems such as ferns (see Figure 4.97).

Mandelbrot identified four properties of fractals: A pattern that repeats under constant iteration, a dimension that is greater than the space they occupy; an infinitely complex structure and a pattern of "self-similarity."

Figure 4.97   Natural examples of fractals

For many years mathematicians flirted with concepts related to fractals, but it was Mandelbrot who created the solid basis for these ideas, illustrating them with remarkable computer-generated visualizations. His foundational *Mandelbrot Set* is a recursive process with infinite iterations of video images that has stirred the popular imagination and can be seen in Figure 4.98.[20] The set possesses a mathematical beauty that has great aesthetic appeal yet arises from the application of basic rules to produce an incredibly complex structure.

There are many mathematical examples of recursive sets, such as the natural numbers which begin with 0 and continue by adding 1 indefinitely. The Fibonacci sequence is also a nice example of recursion which begins with the two numbers 1 and 1, and builds upon itself indefinitely by adding the two previous numbers to produce the next number.

Although Mandelbrot offered a definition of a Fractal as a "rough or fragmented geometric shape that can be split into parts, each of which is approximately a reduced-size copy of the whole," it has not completely satisfied mathematicians who still struggle to accurately define a fractal. Some mathematicians feel that it defies a precise definition. There is agreement, however that fractal patterns possess

---

[20] See the video images at: https://en.wikipedia.org/wiki/Mandelbrot_set and https://en.wikipedia.org/wiki/Fractal

Figure 4.98   An Iteration of the Mandelbrot Set

fractal dimensions which attest to their complexity. The dimension of a fractal is linked to the concept of the *Hausdorff dimension* which is a measure of *roughness* developed in 1918 by the Prussian mathematician Felix Hausdorff (1868–1942). A smooth shape, such as a polyhedron which has a finite number of corners, has a Hausdorff dimension which is an integer. For example, the Hausdorff dimension of a single point is zero, that of a line segment is 1, of a square 2 and a cube 3. For infinite sets of points, the set of natural numbers has Hausdorff dimension zero, the set of points on a plane, dimension 2, and the set of points in three-dimensional space has Hausdorff dimension 3. However, fractals, which can possess infinite "roughness," have *non-integer* dimensions that quantify the changes in their detail as the scale magnifies them.

One type of fractal is a *geometric fractal* that is created from a common geometric figure. One of the most famous geometric fractals is the Sierpiński triangle, shown in Figure 4.99 and is named after the Polish mathematician Waclaw Sierpiński (1882–1969). This triangle did, however, appear as a decorative pattern many years before the work of Sierpiński. On the floors of some Roman churches, dating from the 11th century, a design very similar to the Sierpiński triangle is found in pieces of stone of different sizes that are cut into the desired shapes.

The Sierpinski triangle is constructed by repeated division of an equilateral triangle into smaller and smaller equilateral triangles while removing some of the triangular subsets, as follows:

1. First, divide an equilateral triangle into 4 equilateral triangles and then remove the inner triangle producing 3 equilateral triangles. These are the black triangles in Figure 4.99.

2. Inside each of the 3 black equilateral triangles, whose sides share part of the edges of the original triangle, repeat the process and remove the inner triangle, producing 3 more equilateral triangles within each of the subdivided triangles.

3. Continue the process indefinitely adding 3 equilateral triangles within each subdivided triangle by removing the inner triangle.

Each triangular iteration creates 3 copies of itself and 2 more triangles are produced within each subdivided triangle. The number 3 represents the *self-similar factor*, while 2 is considered the *magnification factor*. The dimension for a geometric fractal is determined as follows: If N = the self-similar factor and S = the magnification factor, the Hausdorff dimension d is given by the formula: $d = \log_S N$.

The Hausdorff dimension of the Sierpinski triangle is then the non-integer: $d = \log_2 3 \approx 1.585$.

There are many types of geometric fractals generated from triangles, squares, cubes etc. and the reader is encouraged to explore the topic further.[21]

---

[21] http://fractalfoundation.org/OFC/OFC-10-3.html Benoit B. Mandelbort, The Fractal Geometry of Nature, New York: W. H. Freeman and Company, 1983.

Figure 4.99   Sierpiński Triangle

Figure 4.100   Maps of England compared

Mandelbrot was inspired by fractals that are not generated from geometric figures but found in nature such as the coastline of England which, when viewed in large scale, appears somewhat straight but as one zooms in, the edge of the coastline appears rougher and rougher revealing twists, inlets, jagged edges, rocks, and other irregularities (see Figure 4.100).

This roughness persists even under a microscope with the uneven perimeter of pebbles, sand and soil. The fractal dimension of such a coastline reveals how "complicated" each self-similar figure is as one zooms in. The English mathematician Lewis Fry Richardson (1881–1953) explored various coastlines to determine their fractal dimension using a formula similar to that given above for geometric fractals.[22] Some of the coastline dimensions that have been calculated are:

South Africa: $d = 1.05$

Australia: $d = 1.13$

Great Britain: $d = 1.25$

Norway: $d = 1.52$

Portugal: $d = 1.14$

The fractal dimension of a line is 1 and the closer the dimension of a coastline is to 1 the less is its departure from a straight line. Fractals have found application in nature, computer imaging, mathematics, 3D modeling, data management and various other areas of technology. As a result, *The Fractal Foundation*[23] has been formed and stated as its mission: "We use the beauty of fractals to inspire interest in science, math and art."

---

[22] http://fractalfoundation.org/OFC/OFC-10-4.html
[23] https://fractalfoundation.org/

Chapter 5

Probability

UNIT 75

The History of Probability

As one might expect, gambling can be considered to have sparked the beginning of probability theory. In the 17th century, two French mathematicians Blaise Pascal (1623–1662) and Pierre de Fermat (1601–1665) were tangling with a problem that was presented to them by Antoine Gombaud, Chevalier de Méré, a French writer. As he had an interest in gambling, he contacted Pascal to help him sort out an apparent contradiction regarding a popular dice game. The game involved tossing a pair of dice 24 times and to decide whether or not to bet even money on the occurrence of at least getting one double-6 amongst these 24 throws. Intuitively, Chevalier de Méré believed that betting on getting a double-6 on 24 throws would be profitable, however his own experience seemed to indicate the opposite. This problem and others posed by Chevalier de Méré to Pascal motivated a continuous correspondence through letters between Pascal and Fermat in which the fundamental principles of probability theory were formulated for the first time. Although a few special problems on games of chance had been solved by some Italian mathematicians in the 15th and 16th centuries, theories of probability were first developed during the correspondence between Pascal and Fermat.

In Chevalier de Méré's dice game, in which players wagered on getting a double-6 on 24 tosses of 2 dice, he correctly reasoned that the chance of getting a double-6 in a single throw of two dice was $\frac{1}{36}$. Then applying his formula, he calculated the chance of success as $24 \times \frac{1}{36} = \frac{2}{3}$. He then decided to wager a large sum of money on getting the double-6s. The more he played the game the more he lost, showing that his speculations were incorrect. This conundrum motivated him in 1654 to contact Pascal who then corresponded with Fermat to make some sense out of *"De Méré's paradox,"* which they solved and that is when the concept of probability, as we know it today, was born.

The probability of not getting a 6 on one toss is $\frac{5}{6}$. The probability of not getting a 6 on 4 tosses is $\left(\frac{5}{6}\right)^4$.

Therefore, the probability of getting at least one "6" in 4 tosses of a single die is:

$$1 - \left(\frac{5}{6}\right)^4 \approx 0.51774691358 \tag{1}$$

which is a bit higher than the probability of at least 1 double-6 in 24 throws of 2 dice,

$$1 - \left(\frac{35}{36}\right)^{24} \approx 0.49140387613 \tag{2}$$

Chevalier de Méré suspected that (1) was higher than (2), but his mathematical skills were not great enough to demonstrate why this should be so. He posed the question to Pascal, who along with Fermat, solved the problem and proved Chevalier de Méré to be correct. In fact, Chevalier de Méré's observation remains true even if 2 dice are thrown 25 times, since the probability of throwing at least 1 double-6 is then $1 - \left(\frac{35}{36}\right)^{25} \approx 0.505531546$. (3)

The genius minds of Pascal and Fermat have in this way been applied to the founding of probability theory.

UNIT 76

The Famous Birthday Problem

There are certain results in the field of probability that lead to rather counterintuitive results. One such situation arises when we try to determine the probability that in a group of people, there will be two people with the same birth date (not including the year). To better understand this situation, let's consider the following example. Suppose you are in a group with about 35 randomly selected people. What do you think the chance (or probability) is that two people in the group would have the same birth date (month and day)? Intuitively, one usually begins to think about the likelihood of 2 people having the same date out of a selection of 365 days (assuming no leap year). Perhaps 2 out of 365? That would only be a probability of $\frac{2}{365} = 0.005479 \approx \frac{1}{2}\%$. This is a minuscule chance.

Suppose we begin by considering the "randomly" selected group of the first 35 presidents of the United States. You may be astonished to learn that there actually are two presidents with the same birth date: The 11th president, James K. Polk (November 2, 1795), and the 29th president, Warren G. Harding (November 2, 1865).

You may be surprised to learn that for a group of 35, the probability that two members will have the same birth date is greater than 8 out of 10, or $\frac{8}{10} = 80\%$.

If you have the opportunity, you may want to try your own experiment by selecting groups of about 35 members to check on date matches. For a group of 30 people, the probability that there will be a match is greater than 7 out of 10, or there is a very good chance that in 7 of these 10 groups there ought to be a match of birth dates. What causes this incredible and unanticipated result? Can this be true? It seems to go against our intuition. To relieve you of your curiosity we will consider the situation in detail. Let's consider a group of 35 people. Consider the question: What do you think the probability is that one selected person matches his own birth date? Clearly it is 1 because it is a certainty and can be written as $\frac{365}{365}$.

The probability that another person does *not* match the first selected person is $\frac{365-1}{365} = \frac{364}{365}$. The probability that a third person does *not* match the first and second persons is $\frac{365-2}{365} = \frac{363}{365}$. The probability of all 35 people *not* having the same birth date is the product of these probabilities: $p = \frac{365}{365} \times \frac{365-1}{365} \times \frac{365-2}{365} \times \cdots \times \frac{365-34}{365}$.

Since the probability (q) that two people in the group <u>have</u> the same birth date and the probability (p) that two people in the group do <u>not</u> have the same birth date is a certainty, the sum of those probabilities must be 1. Thus, $p + q = 1$ or $q = 1 - p$.

Therefore: $q = 1 - \frac{365}{365} \times \frac{365-1}{365} \times \frac{365-2}{365} \times \cdots \times \frac{365-33}{365} \times \frac{365-34}{365} \approx$ 0.8143832388747152. In other words, the probability that there will be a birth date match in a randomly selected group of 35 people is somewhat greater than $\frac{8}{10}$. This is quite unexpected when one considers that there were 365 dates from which to choose. You may want to investigate the nature of the probability function. In Figure 5.1 are a few values to serve as a guide:

Number of people in group	Probability of a birth date match
10	0.1169481777110776
15	0.2529013197636863
20	0.4114383835805799
25	0.5686997039694639
30	0.7063162427192686
35	0.8143832388747152
40	0.891231809817949
45	0.9409758994657749
50	0.9703735795779884
55	0.9862622888164461
60	0.994122660865348
65	0.9976831073124921
70	0.9991595759651571

Figure 5.1

Notice how quickly the probability of almost-certainty is reached. With about 60 people in a group, the chart indicates that it is almost certain (0.99) that two people will have the same birth date — just month and day but not the year.

Furthermore, if you look at the death dates of the first 35 presidents, we can see that two presidents died on March 8th (Millard

Fillmore in 1874 and William H. Taft in 1930) and three presidents died on July 4th (John Adams and Thomas Jefferson in 1826, and James Monroe in 1831). Above all, this astonishing demonstration should serve as an eye-opener about the inadvisability of relying too much on intuition and not on scientific proof.

UNIT 77

The Remarkable Probability in the Game of Craps

You may not be familiar with the game of dice called Craps but it is played in casinos all over the world and was a very popular street game played throughout the 19th and 20th centuries. Craps has a very colorful history and the chance of winning the game has been very cleverly designed mathematically so that it is the only "fair" game in the casino, where the player has the same chance of winning as the house does.[1] Every other game in the casino is designed to favor the house so that they always win in the long run, except with the game of Craps.

The European dice game called Hazard, which possibly dates back to the Middle Ages, was brought to New Orleans from London around 1805 by Bernard de Marigny (1785–1868), a wealthy landowner in Louisiana. The game was significantly modified by de Marigny and evolved into the present game of Craps. The game derives its name from a mispronunciation of the word *crabs*, an abusive term for losing the game by first throwing a 2 or 3 on the dice which are called the craps numbers.

During World War II, craps exploded in popularity, which led to it becoming the dominant casino game in Las Vegas and eventually spreading to casinos worldwide after 2004. There are some 40 different types of bets that one can place each time the game is played, but there are only two that are fair bets. The two fair bets are the "Pass Line" and the "Don't Pass Bar" shown in Figure 5.2.

---

[1]The game of Blackjack could also be considered fair, but only if one has acquired a lot of skill in knowing how to play and bet.

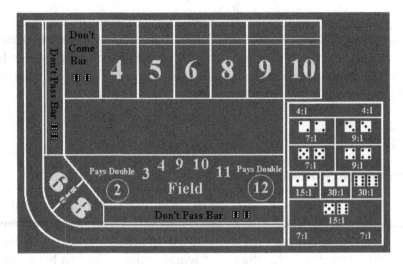

Figure 5.2   Casino craps table

When one puts money down on the "Pass Line" before the player throws the two dice, the bet is that the player will win the game. The "Don't Pass Bar" bets that the player will lose the game. Let us look at the rules of the game, which may seem odd at first, but have been very cleverly devised, as we shall see. There are two dice, and each die has six sides numbered 1 to 6. There are 36 possible outcomes whose totals on the two dice range from 2 to 12 (Figure 5.3). The rules are as follows:

1. The player wins if a 7 or an 11 is thrown on the first toss.

2. The player loses if a total of 2, 3, or 12 is thrown on the first toss.

3. Otherwise, if any of the following totals are thrown on the first toss: 4, 5, 6, 8, 9, or 10, the player must throw the same number again to win the game.

4. If the player throws a 7 before the same number is thrown, the player loses.

5. If the player throws the same number again before a 7 is thrown, the player wins.

Figure 5.3   Throwing the dice

The rules are simple enough but surprisingly yield a fair probability of winning. Associated with each possible way of winning, there is a specific probability. For example, there is a $\frac{1}{6}$ chance of winning by throwing a 7 on the first toss, a $\frac{1}{18}$ chance of winning by throwing an 11 on the first toss, a $\frac{1}{18}$ chance of winning by throwing a 4 or a 10 on the first toss and throwing the same number again, before a 7 is thrown, and so on. When the probabilities of all the possible winning outcomes are calculated and added together, we have the following formula:

$$P(\text{Win}) = \frac{1}{6}(7) + \frac{1}{18}(11) + \frac{1}{18}(4 \text{ or } 10) + \frac{4}{45}(5 \text{ or } 9) + \frac{25}{198}(6 \text{ or } 8)$$

The total of all the terms yields the probability of winning at Craps:

$$P(\text{Win}) = \frac{244}{495} = 0.49292929\ldots$$

This means that the probability of winning is almost $\frac{1}{2}$, which is more than the probability of any other bet on the Craps table (except not winning), or any bet on any other game in the casino. This means that for a very large number of games, very close to 50% of them will be won by the player and close to 50% will be lost. Furthermore and most importantly, the house will pay even money if you bet the Pass Line, or the Don't Pass Bar, and win. For example, if you bet x dollars, you will win x dollars, which means in the long run the house is not favored to win any money on such a bet, so the Pass Line or the Don't Pass Bar is a fair bet for the player. The remarkable aspect of this probability of winning is that we do not know if the rules were arrived at mathematically to make the probability of winning almost 50%, or the rules somehow evolved to this fair arrangement.

The record of consecutive wins is held by Patricia DeMauro, who rolled 154 wins over a period of 4 hours and 18 minutes on May 24, 2009 at the Borgata Casino in Atlantic City, New Jersey.

The following strategy is not meant to encourage gambling but merely to illustrate a betting procedure that can be applied to increase the chance of winning. However, it requires continually doubling your bet when you are losing and reducing it when you are winning, which goes against most person's instincts. The strategy is:

1. Bet a given amount, say $\$x$. If you win, then quit or repeat this bet.

2. If you lose, bet $\$2x$. If you win, you will have won back what you lost on the first bet and you will be ahead $\$x$. Then quit or start over.

3. If you lose again, bet $\$4x$. If you win, you now will have won back what you lost on the first and second bet and you will be ahead $\$x$. Then quit or start over.

4. After each loss, double your bet and if you win you will be ahead $\$x$.

This strategy cannot always work, as clearly the casinos are well aware of it and almost always place a limit on how much one can bet. Good luck if you decide to give it a whirl!

UNIT 78
Some Statements are Neither True nor False

We are led to believe in school that information is either true or false. Our justice system is built upon that premise. However, a spoken language, and today even a video of an event, may not necessarily be an infallible way to communicate true information. In the curious world of mathematical logic there are statements that if true, then they must also be false, and if false then they must also be true. Consider this simple sentence: "This statement is false." Can this be a true statement? If it is, then the statement must be false. If it is false, then the statement must be true. In other words, the statement speaks of its own untruth. This is one version of what is called in Logic, the *Liar Paradox.* A similar paradox is the statement: "In a certain village, the barber shaves everybody who does **not** shave himself." The question then arises: "Who shaves the barber?" If the barber shaves himself, then he is shaving somebody who shaves himself, which contradicts the statement. If the barber does not shave himself, then he should shave himself as the statement claims. This original statement also has an inherent inconsistency as it cannot, be simply true or false. Logicians differ as to what truth value to assign to such a statement. One major consensus considers the statement as being both true and false at the same time, a point of view known as *dialetheism.* Such statements belong to a three-valued *fuzzy logic* system: true, false and true-false. Logic generally assigns a truth value of 0 for false and 1 for true. Fuzzy logic assigns a truth value of any real number from 0 to 1. If T is the truth value of the Liar Paradox, then $T = \text{Not}(T)$, which implies that: $T = 1 - T$, and therefore, it follows that $T = 0.5$. This is one "logical" way to look at it, but Logicians still differ as to how to resolve the paradox.

However, the Liar Paradox is an important example because it relates to one of the most significant mathematical contributions of the 20th century. In 1931, the brilliant Austrian-American logician Kurt Gödel (1906–1978) published his *Incompleteness Theorems,* which had profound implications concerning the structure of arithmetic. As shown in the unit on Euclid's Elements (page 204), the entire system of plane geometry is solidly built upon five axioms, which spawn an enormous number of geometric relationships and theorems. The German mathematician David Hilbert (1862–1943) proposed in the early 20th century to create a firm axiomatic foundation for basic arithmetic, similar to that for plane geometry, with hopes that it would also support higher levels of mathematics, such as real analysis. It was Gödel who dashed Hilbert's hopes when he published his two fundamental *Incompleteness Theorems*. It was his first incompleteness theorem that showed there were inherent inconsistencies in such a substantial system like basic arithmetic as there are also in our spoken language. The theorem proves that in any complex system, there exists a statement, which is true but not provable. A simplified version of Gödel's argument is as follows. The statement, called *the sentence G*, says *"This sentence is not provable."* That is, the sentence cannot be logically shown to have a truth value of 1 or 0 based on the axioms and theorems of the system. Therefore, the sentence G is clearly true, because paradoxically it cannot be proved to be true. In other words, it speaks of its own untruth and is a semantical analog to the Liar Paradox. Gödel points out this relationship in his introduction to the Incompleteness Theorem. His proof of the theorem requires a thorough grounding in logic, but it is essentially based on the same logic as the Liar Paradox. Here is a formal statement of the first Incompleteness theorem:

"Any consistent formal system F within which a certain amount of elementary arithmetic can be carried out is incomplete, that is, there are statements in the language of F which can neither be proved nor disproved in F."

As earth shaking as this theorem appears to be, mathematicians do not really feel threatened by it. A well-known logic professor has made this comment about its importance: "Mathematicians view the theory of arithmetic, and all that which builds upon it, as an enormous edifice. In the depths of its foundation are an enormous array of cobwebs and they function as the underlying supportive logic of the edifice."

UNIT 79

The Pigeonhole Principle

One of the famous (although often neglected) problem-solving techniques is the so-called *pigeonhole principle*. In its simplest form, the pigeonhole principle states that if you have, say, 25 items and 24 boxes to put them in, then at least one box must have more than one item in it. In general terms, in mathematics one would say that *if you have k+1 objects that must be put into k holes, then there will be at least one hole with 2 or more objects in it.*

Here is one illustration of the pigeonhole principle at work.

There are 50 teachers' letterboxes in the school's general office. One day the letter carrier delivers 151 pieces of mail for the teachers. After all the letters have been distributed, one mailbox has more letters than any other mailbox. What is the smallest number of letters it can have?

There is a tendency to "fumble around" aimlessly with this sort of problem, usually not knowing where to start. Sometimes, a guess and test procedure may work here. However, the advisable approach for a problem of this sort is to consider extremes. Naturally, it is possible for one teacher to get all the delivered mail, but this is not guaranteed.

To best assess this situation, consider another extreme case where the mail is as evenly distributed as possible, that is, each letter box gets 3 pieces of mail using up 150 pieces of mail. However, there is still one piece of mail that has not been distributed. Therefore, one teacher would have to receive the 151st piece of mail and will have received the most letters, namely, 4 pieces. By the pigeonhole principle, there were fifty 3-packs of letters for the fifty boxes, and the 151st letter had to be placed into one of those 50 boxes. Thus, the smallest number that one letterbox has more than the others is 4.

We can consider other variations of using the pigeonhole principle such as the following. In a drawer, there are 8 blue socks, 6 green socks and 12 black socks. What is the smallest number of socks that must be taken from the drawer — without looking at them — to be certain of having 2 black socks?

This problem requires a form of the pigeonhole principle, which would have us pick the 8 blue socks and the 6 green socks before a single black sock is selected. The next two picks would have to be black socks. In this situation, it took 16 picks before we were *certain* of getting 2 black socks. Naturally, we might have achieved our goal of getting 2 black socks on our first two tries, but it was not assured and it would have been highly unusual. Even if we picked 10 socks at random, we could not be certain that we had 2 black ones among them.

A simple extension of this problem is illuminating. Now that we know it would take 16 picks to be certain of getting 2 black socks, how many picks would be necessary to be certain of getting 4 black socks? Yes, you only need two more picks, for a total of 18. You have already accounted for the worst case — picking all the socks of the other colors plus 2 black socks with your first 16 picks — so the next two picks can only provide you with 2 black socks.

What is the smallest number of socks that must be taken from the drawer without looking at the socks to be certain of having 2 socks of the same color? At first glance this problem appears to be similar to the problem discussed previously. However, there is a slight difference. In this case, we are looking for a matching pair of socks of *any* color. We now apply extreme-case reasoning, similar to that which we used previously. The worst-case scenario has us picking 1 blue sock, 1 green sock and 1 black sock in our first three picks. Thus, the fourth sock must provide us with a matching pair, regardless of what color it is. The smallest number of socks to guarantee a matching pair is 4.

UNIT 80

The Heads and Tails Conundrum

This lovely little unit will show you how some clever reasoning along with algebraic knowledge of the most elementary kind will help you solve a seemingly "impossibly difficult" problem. Consider the following problem.

> *You are seated at a table in a dark room. On the table there are 12*
> *pennies, 5 of which are heads up and 7 are tails up. (You know where*
> *the coins are, so you can move or flip any coin, but because it is dark*
> *you will not know if the coin you are touching was originally heads up*
> *or tails up.) You are to separate the coins into two piles (possibly flip-*
> *ping some of them) so that when the lights are turned on there will be*
> *an equal number of heads in each pile.*

Your first reaction is "you must be kidding!" "How can anyone do this task without seeing which coins are heads or tails up?" This is where a most clever (yet incredibly simple) use of algebra will be the key to the solution.

Let's "cut to the chase." Here is what you do. (You might actually want to try it with 12 coins.) Separate the coins into 2 piles of 5 and 7 coins each. Then flip over the coins in the smaller pile. Now both piles will have the same number of heads! That's all! You will think this is magic. How did this happen. Well, this is where algebra helps us understand what was actually done.

Let's say that when you separate the coins in the dark room, h heads will end up in the 7-coin pile. Then the other pile, the 5-coin pile, will have $5 - h$ heads and $5 - (5 - h) =$ tails. When you flip all the coins in the smaller pile, the $5 - h$ heads become tails and the h tails become heads. Now each pile contains h heads!

UNIT 81

The Monty Hall Problem

"Let's Make a Deal" was a long-running television game show that featured the following problematic situation. A randomly selected audience member came on stage and was presented with three doors and asked to select one. There are *two donkeys* and *one car* behind these doors. Hopefully, the door selected would contain the car and not one of the other two doors, each of which had a donkey behind it. There was only one wrinkle in this: after the contestant made a selection, the host, Monty Hall, exposed one of the two donkeys behind a door that was not selected (leaving two doors still unopened) and the participant was asked if he/she wanted to stay with the original selection (not yet revealed) or switch to the other unopened door. At this point, to heighten the suspense, the rest of the audience would shout out "stay" or "switch" with seemingly equal frequency. The question is what to do? Does it make a difference? If so, which is the better strategy (i.e. has the greater probability of winning) to use here?

Let us look at this now step-by-step. The result gradually will become clear.

You must try to get the car. You select Door #3 (Figure 5.4)

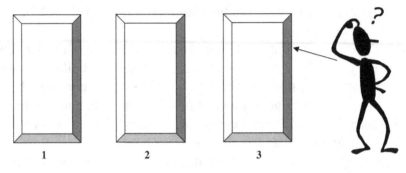

Figure 5.4

Monty Hall opens one of the doors that you *did not* select and exposes a donkey (Figure 5.5).

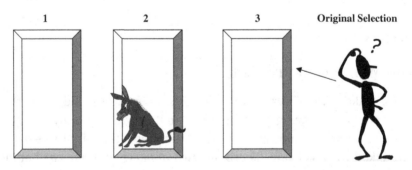

Figure 5.5

He asks: "Do you still want your first-choice door, or do you want to switch to the other closed door"?

To help make a decision, let's consider an *extreme case*:

Suppose there were 1000 doors instead of just three doors (Figure 5.6).

You choose Door #1000. How likely is it that you chose the right door?

Figure 5.6

"Very unlikely" since the probability of getting the right door is $\frac{1}{1000}$. How likely is it that the car is behind one of the other doors? *"Very likely"*: $\frac{999}{1000}$ (see Figure 5.7)

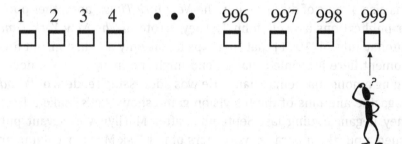

Figure 5.7

These are all *"very likely"* doors!

Monty Hall now opens *all* the doors (2 – 999), *except one* (say, door No. 1) and shows that each one had a donkey (Figure 5.8).

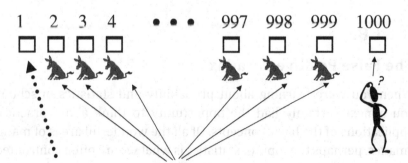

Figure 5.8

A **"very likely"** door is left: Door #1.

We are now ready to answer the question. Which is a better choice:

- Door No. 1000 (*"very unlikely"* door), or
- Door No. 1 (**"very likely"** door)?

The answer is now obvious. We ought to select the "very likely" door, which means "switching" is the better strategy for the audience participant to follow.

In this extreme case, it is much easier to see the best strategy, than had we tried to analyze the situation with just three doors. The principle is the same in either situation.

This problem has caused many arguments in academic circles and was also a topic of discussion in the *New York Times*, and other popular publications as well. John Tierney, wrote in *The New York Times* (Sunday July 21, 1991) that "perhaps it was only an illusion, but for a moment here it seemed that an end might be in sight to the debate raging among mathematicians." He was addressing readers of *Parade* magazine and fans of the television game show "Let's Make a Deal." They began arguing last September after Marilyn vos Savant published a puzzle in Parade. As readers of her 'Ask Marilyn' column are reminded each week, Ms. vos Savant is listed in the Guinness Book of World Records Hall of Fame for "Highest I.Q.," but that credential did not impress the public when she answered this question from a reader. She gave the right answer, but still many mathematicians argued against her.

UNIT 82

The False Positive Paradox

When you were learning about probability and statistics in school, you almost certainly had the opportunity to think about practical applications of the basic concepts. Of all the wonderful areas of mathematics, perhaps the topic of statistics is what we are often confronted with in daily life. It is perhaps not surprising that the simplification of complex data to just a few numbers leads to some loss of information. Nevertheless, taking a closer look at some ideas we might have about such statistical information can yield some big surprises.

An example of such a surprise is the so-called *false positive paradox*. Typically, this imprecise interpretation of probabilities turns up in discussions of testing for diseases.

Imagine that you are going to the doctor for a check-up, and part of this is a test for a specific disease. For the sake of this discussion, let's call it A-Virus. You are told that this recently developed test is the best ever for A-Virus, and 99% accurate. A week later, you get the results of your test, and you are told that you have tested positive. A typical reaction at this point (but not one supported by the numbers, as we will see), is to assume the worst. After all, the test is 99% accurate. So that means that there is a 99% chance that you have A-Virus, right? Well, no. Let's take a closer look at the meaning of the numbers.

When we state that a test is 99% accurate, we are, in fact, not being very precise. Do we mean that 99% of all people with the disease will be diagnosed as having it and 1% will be diagnosed as not having it, even though they do. This is a so-called *false negative* or do we mean that 99% of all healthy people will be correctly diagnosed? Or do we mean both? To simplify things a bit, we can assume here that both of these assumptions are true. (For real tests of this type, these numbers will typically not be the same. The percentage of *false positives* often differs from the percentage of *false negatives*.) So, we assume that 99% of all people taking the test gets a correct result and 1% an incorrect one. It seems obvious that this also means that 99% of all people testing positive will, in fact, have been infected with A-Virus.

In order to see that this reasoning approach needs to be more specific, consider a hypothetical population and check their numbers in detail. Let us assume that we are testing 100,000 people for A-Virus. We need to make some assumption on the number of people in the population that are actually infected. To this end, we assume that 0.1% of the total population actually has A-Virus. This means that 100,000 × 0.1% = 100 people have A-Virus and 100,000 − 100 = 99,900 do not. These assumptions then give us the data collected in the table of Figure 5.9.

Of the 100 infected members of the tested population, 99 tested positive and 1 tested negative, since the test is 99% accurate. However, of the 99,900 healthy individuals, 1% also tested positive. This results in 99,900 × 1% = 999 healthy individuals testing positive, and therefore 99,900 − 999 = 98,901 testing negative.

Adding these numbers, we see that 99 + 999 = 1098 people tested positive. This means that, out of all individuals testing positive, only $\frac{99}{1098} \approx 9.2\%$ are actually infected with the A-Virus. Put another way, even if you test positive, the probability that you are not infected is still about 100% − 9.2% = 90.8%. This is a far cry from the 1% we may have naively assumed.

In fact, the less prevalent the infection is in the total population, the less likely a positive test will reliably reveal infection. If only 0.01% of the population is infected, the percentage drops to $\frac{99}{99+9999} \approx 1\%$. (These numbers result from the same idea as above, but with a tested population of one million.) We see that it is still quite unlikely that you actually have some infection, even if you test positive for it, which is a relatively rare occurrence. This is why more tests are needed under such circumstances. Of course, two false positives in a row become much less likely, and a second test will give us a better idea of the actual situation.

On the other hand, if you test negative, the likelihood that you are free of infection is quite high. Once again, let us consider the numbers in our test population. Of the 98,902 individuals whose tests come back negative, 98,901 are actually free of infection. The probability that someone testing negative is actually not infected is therefore, equal to $\frac{98,901}{98,902} \approx 99.999\%$. This is such a high probability that it is about as close to a sure thing as we can get by statistical methods.

The somewhat revealing nature of this approach is typical for conditional probabilities. Working with such values takes some

	Test Results Positive	Test Results Negative	*Total*
Infected with A-Virus	99	1	100
Not Infected	999	98,901	99,900
Total	1,098	98,902	100,000

Figure 5.9

getting used to, and we need to consider more carefully a complete analysis of the statistics when dealing with such probabilities.

UNIT 83

Fair Division

In the book of Genesis, there is the story of Abraham and Lot, both wealthy men, quarrelling over who should have the land for their herdsmen to use. Abraham makes the following offer to Lot:

"Let there be no strife, I pray thee, between me and between thee and between my herdsmen and between thy herdsmen, for we be brethren. Is not the whole land before thee? Separate thyself, I pray thee; if thou wilt take the left hand, then I will go to the right; or if thou depart to the right hand, I will go to the left."

This has been understood to mean that two parties wishing to divide up some parcel, be it land or cake, that the fairest way to do it, so that both parties feel it is just, is for one to do the division and for the other to choose one of the two pieces. The metaphor of cake-cutting or land partitioning applies to a range of real problems involving the division of something, or settlement of some dispute, so that all parties feel it is a fair process, and no one is envious of any other party.

The problem of fair division between two or more parties is a recently proposed problem that dates to the end of the second world war and was discussed by three Polish mathematicians in 1944. In 1947, it was revealed at a meeting of the Econometric Society in Washington, DC by the Polish mathematician Hugo Steinhaus (1887–1972), on behalf of his two colleagues, that they considered the problem of accomplishing a fair division with the minimum number of cuts. The solution is straightforward when there are two people, as explained above. Around 1960, mathematicians devised an algorithm that can produce an "envy-free" cake division for three players. Further solutions for *n* persons have been proposed since then that are more complex. Because there are several possible algorithms for three people and the solutions significantly increase in complexity as more people are involved, we will consider here the case of three people dividing a metaphorical cake, and the reader is encouraged to research this problem further.[2]

The simplest solution for three people, called the *Selfridge-Conway* procedure, where each person feels it is a fair division and there is no envy, requires that not necessarily the entire cake be divided as follows. Let P1, P2 and P3 be the three people involved in the division process which then proceeds as follows (see Figure 5.10):

1. P1 first cuts the cake into three pieces as equally as possible.

2. P2 carefully trims what appears to be the largest of the three pieces, so that there are at least two equally large pieces. P2 then removes the piece T that was cut off and *it is left out of the division process*. In the event that P2 feels the two largest pieces are equal then no trimming is done.

3. Each player then chooses one of the three remaining pieces in this order: P3, who did no cutting, chooses first, then P2 who made the second cut, and finally P1 who made the first cut (Figure 5.10).

Note that the person who does the cutting does not get to choose until a person who has not cut has chosen. This requires that the person who cuts has to be precise so that the cuts are made as equitably

---

[2] https://en.wikipedia.org/wiki/Envy-free_cake-cutting

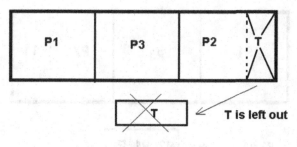

Figure 5.10

as possible. The algorithm then assures that each person has a chance to receive as much as any other person. If there is a piece that is trimmed, each person can expect at least $\frac{1}{4}$ of the total cake as follows: Assuming the worst scenario that P1 cuts the cake where one piece is only $\frac{1}{4}$ of the entire cake and the other two pieces are each approximately $\frac{3}{8}$ of the cake. P2 then trims one of the two larger pieces and discards the cut off piece. P3 then chooses the largest piece, P2 the next largest piece and P1 is left with only $\frac{1}{4}$ of the cake.

A second case for three persons, which divides the entire cake, requires at least five cuts as follows:

1. P1 first cuts the cake into three pieces as equally as possible.

2. P2 then trims what appears to be the largest piece so that there are at least two equal large pieces. The piece that is cut off is put aside but remains in the offing. Call it T1.

   [If P2 feels that the two largest pieces are equal and no trimming is needed, then each player chooses a piece in this order: P3, P2, P1.]

3. P3 now chooses one of the three pieces.

4. P2 now chooses a piece with the stipulation that if P3 did not choose the piece that was trimmed then P2 must choose it. This assures that P2 cuts the piece that was trimmed as equitably as possible.

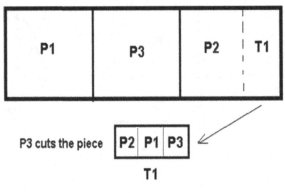

Figure 5.11

5. P1 takes the last piece leaving the cut off piece T1 to be divided. Now between P2 and P3, whoever did not choose the piece that was trimmed, cuts the piece T1 that was put aside into three pieces as equally as possible. Let us assume P2 chose the piece that was trimmed so that P3 cuts T1 into three equal pieces. (The final result is similar if P3 chooses the piece that was trimmed) (see Figure 5.11).

6. P2 then chooses one of the three pieces of the cut off piece T1.

7. P1 then chooses one of the two remaining pieces of T1.

8. P3 then takes the last piece of T1.

This algorithm also assures that each person has a fair chance of getting at least $\frac{1}{3}$ of the cake because the first division into the three largest pieces and the second division of the cut-off piece both follow the same logic. Whoever does the cutting is always the last to choose the piece that was cut, so the cuts must be carefully made to assure that no one receives more than anyone else. This assures each person that they will receive as much as any other person and minimizes any envy.

An extension of fair division is found in *game theory* where the problem of dividing a set of resources among several people who have a claim to them, is done so that each person receives their due share.

The problem of an envy free cake-cutting algorithm for *n* persons was solved in 2016 by two computer scientists, Simon Mackenzie and Haris Aziz,[3] but it involves such an enormous number of cuts as n increases in value, that fair cake-cutting with less steps is still a subject of research in computer science, mathematics, political science and economics.

[3] By Haris Aziz, Simon Mackenzie Communications of the ACM, April 2020, Vol. 63, No. 4, pp. 119–126.

Chapter 6

A Potpourri of Mathematical Topics

UNIT 84

Logical Thinking

When a problem is posed that at first looks a bit daunting, and then a solution — one easily understood — is presented, we often wonder why we didn't think of that simple solution ourselves. It is exactly these problems that should have a dramatic effect on us, and may help us solve future problems. Here is one such situation.

On a shelf in Danny's basement, there are 3 boxes. One contains only nickels, one contains only dimes, and one contains a mixture of nickels and dimes. The three labels, "nickels," "dimes," and "mixed" fell off, and were all put back on the wrong boxes. Without looking, Danny can select one coin from one of the mislabeled boxes and then correctly label all 3 boxes. From which box should Danny select the coin?

One may reason that the "symmetry" of the problem situation dictates that whatever we can say about the box mislabeled "nickels" could just as well have been said about the box mislabeled "dimes." Thus, if Danny chooses a coin from either of these boxes, the results would be the same.

You should, therefore, concentrate your investigations on what happens if we choose from the box mislabeled "mixed." Suppose Danny selects a nickel from the "mixed" box. Since this box is mislabeled, it cannot be the mixed box and must be, in reality, the nickel box. Since the box marked dimes cannot really be dimes, it must be the "mixed" box. This leaves the third box to be the dimes box. You are probably thinking how simple the problem is — now that you have the solution. It does demonstrate a certain beauty of logical thinking.

UNIT 85

More Mathematical Conundrums

Some conundrums in the form of paradoxes have been shown in the unit on infinity (page 48), which presents the intriguing Xeno's paradox, and in the unit on true-false statements (page 293), where the liar paradox and the barber paradox are presented. However, there are other paradoxes in mathematics that can be fun to explore and serve instructional functions. A paradox, by definition, is a statement that leads to a self-contradiction or a logically unacceptable conclusion, but could possibly be shown to be true or false, with further examination.

Here is a classic mathematical conundrum which "proves" that the weight of an elephant equals the weight of a flea. Let E = the weight of an elephant and f = the weight of a flea. The average weight A of the elephant and the flea is then:

$$A = \frac{E + f}{2}$$

Multiply both sides by 2 to clear the fraction: $2A = E + f$

Solve the equation for E and for $-f$: $E = 2A - f$ and $E - 2A = -f$

Multiply the two left sides and the two right sides of each equation, which preserves the equality:

$$(E)(E - 2A) = (2A - f)(-f)$$

Multiply out the parentheses: $E^2 - 2AE = -2Af + f^2$

Reverse the right side of the equation: $E^2 - 2AE = f^2 - 2Af$

Add A^2 to both sides: $E^2 - 2AE + A^2 = f^2 - 2Af + A^2$

Factor each side of the equation: $(E - A)^2 = (f - A)^2$

Take the square root of each side: $E - A = f - A$

We then have: $E = f$

So, the elephant weighs as much as the flea? Clearly this cannot be. Can you find the error? Try substituting values for E and f and see what happens before we take the square root of each side. The error is explained at the end of this unit.

Here is another conundrum. Three friends check into a hotel and the total bill is $300. Each person pays $100, however later the manager discovers that they are having a special for two or more people checking in and there is a discount of $50. He then gives his assistant $50 to return to the three people. The assistant realizes that the $50 cannot be evenly divided among the three people and decides to keep $20 as a tip and returns only $30 to the three people. Now each person receives back $10 and has, therefore, paid $90 for the room whose total cost is now $270. The assistant has kept $20, which, added to the $270 spent by the guests, means that the total money paid is $290. But the guests originally handed over $300 to the hotel for the room, so what happened to the other $10? Here lies the paradox, as only one answer, $290 or $300 can be correct. As in the first conundrum there is a flaw in the logic. Can you explain the contradiction? The contradiction is revealed below.

In the first conundrum the error lies in these steps:

$$(E - A)^2 = (f - A)^2$$

Take the square root of each side: $E - A = f - A$

If A is the average of E and f, then the weight of the elephant E is clearly greater than A and $E - A$ is positive, whereas the weight of the flea f is less than A, which means $f - A$ is negative. Therefore, the positive square root of $(f - A)^2$ is not $f - A$ which is a negative number, and the result cannot be true. The positive square root of $(f - A)^2$ is actually $A - f$.

In the second conundrum, $300 has clearly been dispersed, where $250 has been paid to the hotel, $20 was taken by the assistant and $30 given back to the guests, so there is no missing $10. The flaw in the reasoning lies in assuming that the hotel was paid $90 from each guest for the room totaling $270, which is not true since the hotel only received $250. By only adding up what the guests paid and what the assistant took does not take into account what the hotel received and, therefore is not a correct sum of the money that was dispersed.

Logic certainly plays an important part in finding errors and seeking the truth, not just in mathematics, but in one's everyday life.

UNIT 86

Mistakes in Mathematics

We tend to frown at the notion of making mistakes in mathematics. Typically, mistakes are found on students' test papers and even sometimes the math teacher will make mistakes in class — often with the excuse "I just wanted to see if you were paying attention." However, there are mistakes that are quite surprising such as the classic one involving the famous Popeye cartoon.

We have been shown through the Popeye comics that spinach brought him extra strength. This is largely a result of some misunderstandings, or perhaps mistakes, about the iron contents in spinach. Spinach has approximately 3.5mg of iron per 100g of spinach, while in its cooked version, it contains about 2mg, which turns out to be a lot less than bread, meat, or fish. This misunderstanding of spinach's iron value stems back to the 1930s where there was a printing mistake. A decimal point was mistakenly moved one place to the right resulting in the value of iron in spinach to be ten times the correct amount, namely, 35mg per 100g of fresh spinach.[1] With Popeye's reinforcement, many children grew up imagining spinach as an instant source of power.

The Hungarian mathematician George Pólya (1887–1985) was known to have said that *"Geometry is the science of correct reasoning on incorrect figures."* We will demonstrate below that making conclusions based on "incorrect" figures can lead us to impossible results. You may find the demonstration of proving something that as absurd as proving that all triangles are isosceles to be either frustrating or enchanting, depending on your disposition. Nevertheless, follow each statement of the "proof" and see if you can detect the mistake. The error is one that Euclid in his book *Elements* would not have been able to resolve because of a lack of a definition.

Geometric Mistake

The Fallacy: *Any scalene triangle (a triangle with three unequal sides) is isosceles (a triangle having two equal sides).*

To prove that scalene $\triangle ABC$ is isosceles, we must draw a few auxiliary line segments. Draw the bisector of $\angle C$ and the perpendicular bisector of AB. From their point of intersection, G, draw perpendiculars to AC and CB, meeting them at points D and F, respectively. There are four possibilities for the above description for various scalene triangles. We will consider each of these, beginning with Figure 6.1, where CG and GE meet *inside* the triangle.

[1]See: Sutton, Mike. "Spinach, Iron, and Popeye: Ironic lessons from biochemistry and history on the importance of healthy eating, healthy skepticism and adequate citation," Internet Journal of Criminology 2010, pp. 1–34: http://www.internetjournalof-criminology.com/Sutton_Spinach_Iron_and_Popeye_March_2010.pdf

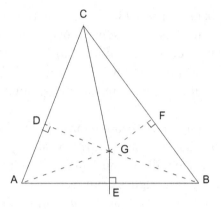

Figure 6.1

The second case is where *CG* and *GE* meet on side *AB*, which is shown in Figure 6.2.

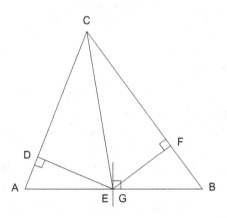

Figure 6.2

The third possibility is where *CG* and *GE* meet outside the triangle, but the perpendiculars *CG* and *GE* intersect the sides *AC* and *CB*, which is shown in Figure 6.3.

The last possibility for this configuration, which is shown in Figure 6.4 is where *CG* and *GE* meet outside the triangle, but the perpendiculars *GD* and *GF* meet the extensions of sides *CA* and *CB* outside the triangle.

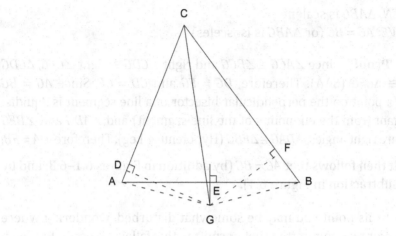

Figure 6.3

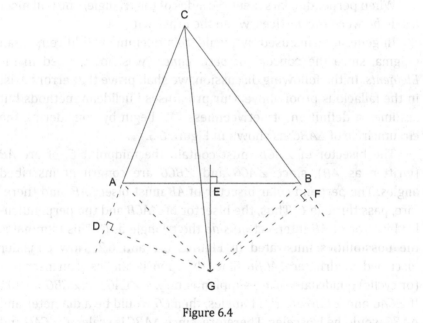

Figure 6.4

The "proof" of the fallacy can be done with any of these four figures.

GIVEN: $\triangle ABC$ is scalene.
PROVE: $AC = BC$ (or $\triangle ABC$ is isosceles)

"Proof": Since $\angle ACG \cong \angle BCG$ and right $\angle CDG \cong$ right $\angle CFG$, $\triangle CDG \cong \triangle CFG$ (SAA). Therefore, $DG = FG$ and $CD = CF$. Since $AG = BG$ (a point on the perpendicular bisector of a line segment is equidistant from the endpoints of the line segment) and $\angle ADG$ and $\angle BFG$ are right angles, $\triangle DAG \cong \triangle FBG$ (Hypotenuse-Leg). Therefore, $DA = FB$.

It then follows that $AC = BC$ (by addition in Figures 6.1–6.3; and by subtraction in Figure 6.4).

At this point you may be somewhat disturbed, wondering where the error was committed that permitted this fallacy to occur. By rigorous construction, you will find a subtle error in the figures:

The point G *must* be outside the triangle.

When perpendiculars meet the sides of the triangle, one will meet a side *between* the vertices, while the other will not.

In general terms used by Euclid, this dilemma would remain an enigma, since the concept of *betweenness* was not defined in his *Elements*. In the following discussion, we shall prove that errors exist in the fallacious proof above. Our proof uses Euclidean methods but assumes a definition of betweenness. We begin by considering the circumcircle of $\triangle ABC$, as shown in Figure 6.5.

The bisector of $\angle ACB$ must contain the midpoint G, of arc AB (written as $\overset{\frown}{AB}$), since $\angle ACG$ and $\angle BCG$ are congruent inscribed angles. The perpendicular bisector of AB must bisect $\overset{\frown}{AB}$, and therefore, pass through G. Thus, the bisector of $\angle ACB$ and the perpendicular bisector of AB intersect *outside* the triangle at G. This eliminates the possibilities illustrated in Figures 6.1 and 6.2. Now consider inscribed quadrilateral $ACBG$. Since the opposite angles of an inscribed (or cyclic) quadrilateral are supplementary, $m\angle CAG + m\angle CBG = 180°$. If $\angle CAG$ and $\angle CBG$ are right angles, then CG would be a diameter and $\triangle ABC$ would be isosceles. Therefore, since $\triangle ABC$ is scalene, $\angle CAG$ and $\angle CBG$ are not right angles. In this case, one must be acute and the other obtuse. Suppose $\angle CBG$ is acute and $\angle CAG$ is obtuse. Then in $\triangle CBG$ the altitude on CB must be *inside* the triangle, while in obtuse $\triangle CAG$, the altitude on AC must be *outside* the triangle. This is usually

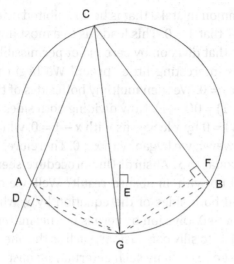

Figure 6.5

readily accepted without proof but can be easily proved. The fact that the one, and only one, of the perpendiculars intersects a side of the triangle between the vertices destroys the fallacious "proof."

This rather thorough discussion of this famous geometric fallacy will give you a real appreciation of the precision of geometry. Beyond the entertainment of this unit, it is very instructional, for it gives reason for defining the location of points, something very often neglected.

Algebraic Mistake

Mistakes can also be made in algebra. For example, students are taught in elementary algebra that $\sqrt{ab} = \sqrt{a} \cdot \sqrt{b}$. However, what is typically omitted is that this is only true when a and b are positive numbers. When we consider what happens when a and b are negative numbers, we are faced with a dilemma. Follow along and you will see what happens when a and b are negative. Let us take a and b, each to be equal to -1. Then for $\sqrt{ab} = \sqrt{a} \cdot \sqrt{b}$, we get the following: $\sqrt{-1}\sqrt{-1} = \sqrt{(-1)(-1)} = \sqrt{+1} = +1$. However, we could also get the following: $\sqrt{-1}\sqrt{-1} = (\sqrt{-1})^2 = -1$, which then leads us to conclude a mistake, since $+1 \neq -1$. The error lies in the fact that $\sqrt{ab} = \sqrt{a} \cdot \sqrt{b}$ does not hold for negative numbers.

Another common mistake that is best exhibited by considering the following "proof" that $1 = 0$. This leads us to a most important mathematical concept: that division by zero is not permissible. Follow along as we show this interesting little "proof." We begin with our given information that $x = 0$. We then multiply both sides of this equation by $x - 1$ to get $x(x - 1) = 0(x - 1)$. Now dividing both sides of the resulting equation, $x(x - 1) = 0$ by x leaves us with $x - 1 = 0$, which, in turn, tells us that $x = 1$. However, we began with $x = 0$. Therefore, 1 must equal 0, since both are equal to x. Absurd! Our procedure seemed correct so why did we end up with an absurd result? Well, we divided by zero when we divided both sides of the equation by x, which we did not realize was equal to 0. Division by zero is not permitted in mathematics, as it will lead us to silly conclusions, such as the one we just showed above. This is just one of many such entertaining mistakes that give us a more genuine understanding of the "rules" of mathematics.

Geometric Mistake

Another geometric mistake can be seen with the following "proof" that a triangle in a plane can have more than 180°. Follow along as we proceed through this so-called "proof" and see if you can find where the mistake lies. With two intersecting circles of different or the same sizes, we will draw the diameters from one of their points of intersection, point P in Figure 6.6, and then connect the other ends of the diameters.

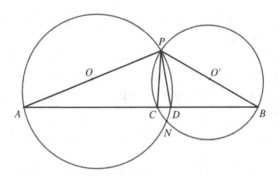

Figure 6.6

The endpoints of diameters AP and BP are connected by line AB, which intersects circle O at point D and circle O' at point C, we find that $\angle ADP$ is inscribed in semicircle PNA, as is $\angle BCP$ inscribed in semicircle PNB, thus making them each right angles. We then have a dilemma: triangle CPD has two right angles! This is impossible. Therefore, there must be a mistake somewhere.

The concern with the omission of the concept of betweenness in Euclid's work could lead us to this dilemma. When we draw this figure correctly, we find that the angle CPD must equal 0, since a triangle cannot have more than 180°. That would make the triangle CPD nonexistent. Figure 6.7 shows the correct drawing of this situation.

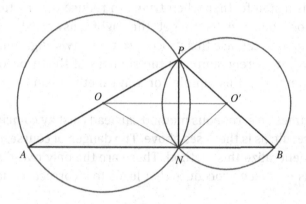

Figure 6.7

In Figure 6.7, we can easily show that $\triangle POO' \cong \triangle NOO'$, and then $\angle POO' \cong \angle NOO'$. Because $\angle PON = \angle A + \angle ANO$, and $\angle ANO = \angle NOO'$ (alternate-interior angles) we have $\angle POO' = \angle A$, and then AN is parallel to OO'. The same argument can be made for circle O' to get BN is parallel to OO'. Since each of the 2 line segments AN and BN are parallel to OO', they must in fact be one line ANB. This proves that the diagram in Figure 6.7 is correct and the diagram in Figure 6.6 is not.

Arithmetic Mistake

There are times when mistakes are simply silly and at the same time convincing for the uninformed. Consider for example, a mistake that has been sometimes referred to as a "howler." It is just the kind of mistake that can surely bring us to wonder! Consider the following, where we are asked to reduce the fraction $\frac{16}{64}$ and we simply cross out the 6's: $\frac{16}{64} = \frac{1\!\!\!/6}{6\!\!\!/4} = \frac{1}{4}$, which, strangely enough, gives us the correct result. We can apply this method to other fractions as well:

To reduce the fraction $\frac{26}{65}$, we simply cross out the 6s to once again get the right answer: $\frac{26}{65} = \frac{2\!\!\!/6}{6\!\!\!/5} = \frac{2}{5}$. Another example, where this strange procedure works is when we wish to reduce the fraction $\frac{19}{95}$, we simply cross out the 9s to get the right answer: $\frac{19}{95} = \frac{1\!\!\!/9}{9\!\!\!/5} = \frac{1}{5}$. It is also possible to take this a step further when trying to reduce the fraction $\frac{49}{98}$, we simply cross out the 9s to get the right answer: $\frac{49}{98} = \frac{4\!\!\!/9}{9\!\!\!/8} = \frac{4}{8} = \frac{1}{2}$. Even though one can use this procedure for all two-digit multiples of $11(\frac{1\!\!\!/1}{1\!\!\!/1}, \frac{22}{22}, \ldots)$, it merely supports the silliness of this procedure. One then wonders why this simple (or silly) method cannot always be used.

Sometimes an erroneous method can lead (just by coincidence) to a correct result, as is the cases above. The danger, of course, is that we must not generalize this method. These are the only two-digit examples of this *mistaken* procedure that leads to a correct reduction of a fraction.[2]

An arithmetic explanation for why this works can be seen from the following calculation: $\frac{16}{64} = \frac{1\cdot10+6}{10\cdot6+4} = \frac{6\cdot\frac{16}{6}}{6\cdot\frac{64}{6}} = \frac{6\cdot\frac{8}{3}}{6\cdot\frac{32}{3}} = \frac{8}{32} = \frac{1}{4}$; therefore, $\frac{16}{64} = \frac{1\cdot10+6}{10\cdot6+4} = \frac{1}{4}$.

There are many examples that could be considered to exhibit mistakes in mathematics. They are worth pursuing, since in almost every case they shed light on mathematical principles that must be adhered to. A book that can provide a wide variety of interesting mistakes is *Magnificent Mistakes in Mathematics* by A.S. Posamentier and I. Lehmann (Prometheus Books, 2013), which offers many curious

---

[2] Ogilvy, C. Stanley; Anderson, John T.: *Excursions in Number Theory*. New York, Oxford University Press, 1966, p. 86.

mistakes in arithmetic, algebra, geometry, probability, and statistics, each of which provides greater insight into mathematics.

UNIT 87

The Hands of an Analog Clock

The clock can be an interesting source of mathematical applications. A common application of mathematics in the school curriculum is about uniform motion involving cars and trains. However, a clock's hands also travel at uniform speed and allow us to enjoy some unusual aspects of time travel, however this time with the hands of the clock.

Begin by determining the *exact time* that the hands of a clock will overlap after 4:00 o'clock. Your first reaction to the solution to this problem is likely to be that the answer is simply 4:20. That does not take into account that the hour hand moves uniformly while the minute hand also moves uniformly but faster. With this in mind, the astute reader will begin to estimate the answer to be between 4:21 and 4:22. You should realize that the hour hand moves through an interval between minute markers every 12 minutes. Therefore, it will leave the interval 4:21–4:22 at 4:24. This, however, does not answer the original question about the exact time of this overlap.

We will provide a "trick" to deal better with this situation. Realizing that this first guess of 4:20 is not the correct answer, since the hour hand does not remain stationary and moves when the minute hand moves, the trick is to simply multiply the 20 (the wrong answer) by $\frac{12}{11}$ to get $21\frac{9}{11}$, which yields the correct answer: $4:21\frac{9}{11}$.

One way to understand the movement of the hands of a clock is by considering the hands traveling independently around the clock at uniform speeds. The minute markings on the clock (from now on referred to as "markers") will serve to denote distance as well as time. An analogy should be drawn here to the "uniform motion" of automobiles (a popular topic for verbal problems in an elementary algebra course). A problem involving a fast automobile overtaking a slower one would be analogous. Experience has shown that the analogy might be helpful to compare the distance necessary for a car traveling at 60 m.p.h. to overtake a car with a head start of 20 miles and traveling at 5 m.p.h.

Now consider 4 o'clock as the initial time on the clock. Our problem will be to determine exactly when the minute hand will overtake the hour hand after 4 o'clock. Consider the speed of the hour hand to be r, then the speed of the minute hand must be $12r$. We seek the distance, measured by the number of markers traveled, that the minute hand must travel to overtake the hour hand. Let us refer to this distance as d markers. Hence the distance that the hour hand travels is $d - 20$ markers, since it has a 20-marker head start over the minute hand (see Figure 6.8).

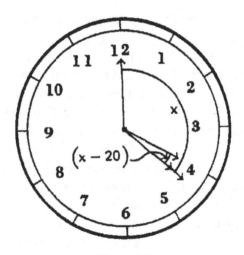

Figure 6.8

For this to take place, the times required for the minute hand, $\frac{d}{12r}$, and for the hour hand, $\frac{d-20}{r}$, are the same. Therefore, $\frac{d}{12r} = \frac{d-20}{r}$, and $d = \frac{12}{11} \times 20 = 21\frac{9}{11}$. Thus, the minute hand will overtake the hour hand at exactly $4:21\frac{9}{11}$.

Consider the expression $d = \frac{12}{11} \times 20$. The quantity 20 is the number of markers that the minute hand had to travel to get to the desired position, if we assume the hour hand remained stationary. However, quite obviously, the hour hand does not remain stationary.

Hence, we must multiply this quantity by $\frac{12}{11}$, since the minute hand must travel $\frac{12}{11}$ as far. Let us refer to this fraction, $\frac{12}{11}$, as the correction factor.

To familiarize yourself with the use of the correction factor, choose some short and simple problems. For example, you may seek to find the exact time when the hands of a clock overlap between 7 and 8 o'clock. Here you would first determine how far the minute hand would have to travel from the "12" position to the position of the hour hand, assuming again that the hour hand remains stationary. Then by multiplying the number of markers, 35, by the *correction factor*, $\frac{12}{11}$, you will obtain the exact time, $7:38\frac{2}{11}$, that the hands will overlap.

To enhance your understanding of this new procedure, consider a person checking a wristwatch against an electric clock and noticing that the hands on the wristwatch overlap every 65 minutes (as measured by the electric clock). Is the wristwatch fast, slow, or accurate? You may wish to consider the problem in the following way. At 12 o'clock the hands of a clock overlap exactly. Using the previously described method we find that the hands will again overlap at exactly $1:05\frac{5}{11}$, and then again at exactly $2:10\frac{10}{11}$, and again at exactly $3:16\frac{4}{11}$ and so on. Each time there is an interval of $65\frac{5}{11}$ minutes between overlapping positions. Hence, the person's watch is inaccurate by $\frac{5}{11}$ of a minute. Can you now determine if the wristwatch is fast or slow?

There are many other interesting and sometimes rather difficult, problems made simple by this "correction factor." You may very easily pose your own problems. For example, you may wish to find the exact times when the hands of a clock will be perpendicular (or form a straight angle) between, say, 8 and 9 o'clock. Again, you would try to determine the number of markers that the minute hand would have to travel from the "12" position until it forms the desired angle with the stationary hour hand. Then multiply this number by the correction factor, $\frac{12}{11}$, to obtain the exact actual time. That is, to find the exact time that the hands of a clock are *first* perpendicular between 8 and 9 o'clock, determine the desired position of the minute hand when the hour hand remains stationary (here, on the 25-minute marker).

Then, multiply 25 by $\frac{12}{11}$ to get $8:27\frac{3}{11}$, the exact time when the hands are *first* perpendicular after 8 o'clock.

For those who want to look at this issue from a non-algebraic viewpoint, you could justify the $\frac{12}{11}$ correction factor for the interval between overlaps in the following way: Think of the hands of a clock at noon. During the next 12 hours (i.e. until the hands reach the same position at midnight) the hour hand makes one revolution, the minute hand makes 12 revolutions, and the minute hand coincides with the hour hand 11 times (including midnight, but not noon, starting just after the hands separate at noon). Since each hand rotates at a uniform rate, the hands overlap each $\frac{12}{11}$ of an hour, or $65\frac{5}{11}$minutes. This can be extended to other situations. You should derive a great sense of achievement and enjoyment as a result of employing this simple procedure to solve what usually appears to be a very difficult clock problem.

UNIT 88

Pythagorean Triples

Perhaps the best-known relationship in school mathematics is the Pythagorean theorem that everybody seems to remember as $a^2 + b^2 = c^2$, which refers to the side lengths of a right triangle. The most common is the right triangle with side lengths of 3, 4, 5. The question then arises: How can we generate primitive Pythagorean triples (Primitive Pythagorean triples are those that have no common factor among the 3 lengths.) or as important, how can we obtain all Pythagorean triples? That is, is there a formula for achieving this goal? One such formula, attributed to Euclid, for integers m and n, generates values of a, b, and c, where $a^2 + b^2 = c^2$, as follows: $a = m^2 - n^2$, $b = 2mn$, $c = m^2 + n^2$ (assuming $m > n$).

We can easily show that this formula will always yield a Pythagorean triple. If we square each of the terms and then show that the sum of the first two squares is equal to the third square.

$a^2 = (m^2 - n^2)^2$, $b^2 = (2mn)^2$, $c^2 = (m^2 + n^2)^2$. We will do this simple algebraic task by showing that the sum $a^2 + b^2$ is actually equal to c^2.

$a^2 + b^2 = (m^2 - n^2)^2 + (2mn)^2 = m^4 - 2m^2n^2 + n^4 + 4m^2n^2 = m^4 + 2m^2n^2 + n^4$ and $c^2 = (m^2 + n^2)^2 = m^4 + 2m^2n^2 + n^4$ Therefore, $a^2 + b^2 = c^2$.

We can apply Euclid's Formula to gain an insight into properties of Pythagorean triples.

When we insert some values of m and n, as in the table of Figure 6.9, we should notice a pattern that would tell us when the triple will be primitive — which, you will recall, is when the largest common factor of the three numbers is 1 — and also discover some other possible patterns.

m	n	$m^2 - n^2$	$2mn$	$m^2 + n^2$	Pythagorean triple	primitive
2	1	3	4	5	(3, 4, 5)	Yes
3	1	8	6	10	(6, 8, 10)	No
3	2	5	12	13	(5, 12, 13)	Yes
4	1	15	8	17	(8, 15, 17)	Yes
4	2	12	16	20	(12, 16, 20)	No
4	3	7	24	25	(7, 24, 25)	Yes
5	1	24	10	26	(10, 24, 26)	No
5	2	21	20	29	(20, 21, 29)	Yes
5	3	16	30	34	(16, 30, 34)	No
5	4	9	40	41	(9, 40, 41)	Yes
6	1	35	12	37	(12, 35, 37)	Yes
6	2	32	24	40	(24, 32, 40)	No
6	3	27	36	45	(27, 36, 45)	No
6	4	20	48	52	(20, 48, 52)	No
6	5	11	60	61	(11, 60, 61)	Yes
7	1	48	14	50	(14, 48, 50)	No
7	2	45	28	53	(28, 45, 53)	Yes
7	3	40	42	58	(40, 42, 58)	No
7	4	33	56	65	(33, 56, 65)	Yes
7	5	24	70	74	(24, 70, 74)	No
7	6	13	84	85	(13, 84, 85)	Yes
8	1	63	16	65	(16, 63, 65)	Yes
8	2	60	32	68	(32, 60, 68)	No
8	3	55	48	73	(48, 55, 73)	Yes
8	4	48	64	80	(48, 64, 80)	No
8	5	39	80	89	(39, 80, 89)	Yes
8	6	28	96	100	(28, 96, 100)	No
8	7	15	112	113	(15, 112, 113)	Yes

Figure 6.9   Using Euclid's formula to generate pythagorean triples

An inspection of the triples in Figure 6.9, leads to the following conjectures — which, indeed, can be proved. For example, Euclid's formula $a = m^2 - n^2, b = 2mn, c = m^2 + n^2$ will yield *primitive* Pythagorean triples only when m and n are relatively prime — that is, when they have no common factor other than 1 — and *exactly one* of these must be an even number, with $m > n$.

One can even show the fundamental result — that *all* primitive Pythagorean triples can be obtained with Euclid's formula: Every primitive Pythagorean triple can be written as $(m^2 - n^2, 2mn, m^2 + n^2)$ with unique natural numbers m and n, which are relatively prime, $m > n$, and $m - n$ is odd.

A truly clever approach to generate Pythagorean triples was developed by the German mathematician Michael Stifel (1487–1567). He created a sequence of mixed numbers of the following form: $1 + \frac{1}{3}, 2 + \frac{2}{5}, 3 + \frac{3}{7}, 4 + \frac{4}{9}, 5 + \frac{5}{11}, 6 + \frac{6}{13}, 7 + \frac{7}{15}, \cdots$.

This sequence has a nice pattern that will be easy to remember. The whole number parts of the above mixed numbers are simply the natural numbers in order, the numerators of the fractions are the same number as the whole number, and the denominators of the fractions are consecutive odd numbers, beginning with the number 3. When we convert each of the mixed numbers in this sequence to an improper fraction, they will produce the first two members of a Pythagorean triple. For example, if we take the sixth term of this sequence, $6 + \frac{6}{13} = \frac{84}{13}$, we have the first two members of the Pythagorean triple $(13, 84, c)$. Then, to get the third member, we simply obtain c in the following manner: $c^2 = 13^2 + 84^2 = 169 + 7{,}056 = 7{,}225$, and then take the square root of 7,225 to get 85. Thus, the complete Pythagorean triple is $(13, 84, 85)$.

What would appear to be a magic trick, is, in fact, easy to explain: We can write the nth number in Stifel's sequence as $n + \frac{n}{2n+1} = \frac{n(2n+1)+n}{2n+1} = \frac{2n^2+2n}{2n+1} = \frac{2n(n+1)}{2n+1}, n = 1, 2, 3, \ldots$.

This then can result in a Pythagorean triple: $a_n = 2n + 1$, $b_n = 2n(n + 1)$, and $c_n = b_n + 1$. Thus, we see that the Pythagorean triples generated with this technique produces triples where the larger two sides differ by 1, as opposed to Euclid's method which generates all Pythagorean triples.

UNIT 89
Fermat's Last Theorem

Much has been written about one of the oldest relationships in the history of mathematics, the Pythagorean Theorem, which is several thousand years older than the Bible. Several articles in this book discuss aspects of the Pythagorean Theorem, which happens to be one of the few facts most students remember from high school mathematics, that is: $a^2 + b^2 = c^2$. Here is a particular fascinating fact about this famous theorem discussed in the previous unit *Pythagorean Triples* (page 324). While there exists an infinite number of sets of three natural numbers that satisfy the Pythagorean Theorem such as: $3^2 + 4^2 = 5^2$, $5^2 + 12^2 = 13^2$, $8^2 + 15^2 = 17^2$, ... etc., there does not exist *any* combination of three natural numbers that satisfies the relationship: $x^n + y^n = z^n$ for any value of $n > 2$. An easy idea to understand, but a confounding result that is somewhat astonishing mathematically.

The first person to claim a proof that no natural number $n > 2$ can be found to satisfy $x^n + y^n = z^n$ was the famous French lawyer and mathematician Pierre de Fermat (1607–1665).

**Pierre de Fermat
(1601-1665)**

Figure 6.10

Around 1637, Fermat first stated this proposition as a theorem in the margin of the journal *Arithmetica*. However, it is very curious that no proof was given in the journal. Instead, there was a statement in the margin written by Fermat — and discovered 30 years after his death — that the proof he claimed to have was too large to fit in the margin. Figure 6.11 shows a copy of the journal containing Fermat's Last Theorem published posthumously by his son in 1670. Fermat's reputation was definitely credible, being supported by other statements given without proofs that were eventually proven, so this claim was subject to question for many years.

The 1670 edition of **Diophantus'** *Arithmetica* includes Fermat's "Last Theorem" (*Observatio Domini Petri de Fermat*).

Figure 6.11

As the years progressed after Fermat's death, no proof was forth-coming, although no one had found a contradiction to the theorem. It, therefore, became a challenging conjecture and resulted in more unsuccessful attempts at a proof than for any other theorem, becoming one of the most famous mathematical ideas and appeared in the *Guinness Book of World Records* as the "most difficult mathematical problem." People questioned whether Fermat actually ever had a successful proof of the theorem, although one of his proofs for $n = 4$ has been discovered.

Slowly, proofs emerged for Fermat's Last Theorem over the next two hundred years (1637–1839) for the prime values of $n = 3, 5$, and 7. In 1850, the German mathematician Ernst Kummer (1810–1893) proved the theorem for a specific class of primes he called *regular primes*, whose definition is too complex to include here. With the advent of computers and more sophisticated approaches, mathematicians were able to build upon Kummer's proof and extend it to all prime exponents up to four million, but a complete proof seemed beyond reach with current methods.

However, after 358 years of considerable effort, the first definitive proof was arrived at by the Cambridge mathematician Andrew Wiles (1953–), who secretly worked alone for six years, disguising his efforts by only releasing small aspects of his work and confiding only in his wife.

Wiles first revealed his proof on June 23, 1993 at a Cambridge lecture, however in September 1993 the proof was found to have an error. After input from a friend Nick Katz, he announced the corrected proof one year later in September 1994 at the Isaac Newton Institute in Cambridge, England. Wiles called this "the most important moment of [his] working life" because he had an illuminating discovery that provided the correction to the proof. He published the proof in 1995 and was later awarded the *Abel Prize* in 2016, "for his stunning proof of Fermat's Last Theorem". The Abel prize, a prize awarded annually by the King of Norway, is named after the outstanding Norwegian mathematician Niels Henrik Abel (1802–1829) and is designed to be the equivalent of the Nobel Prize. Alfred Nobel (1833–1896), a Swedish chemist, chose not to create an award in mathematics possibly because the purpose of the Nobel foundation was to recognize scientific work that could benefit humanity, and it was felt during his time that mathematics did not serve that purpose directly.[3]

UNIT 90

Pure Mathematics and Prime Numbers

The famous British mathematician Godfrey H. Hardy (1877–1947) wrote a treatise in 1940 called *"A Mathematician's Apology"* in which he defends and justifies pursuing mathematics strictly for its intrinsic value and inherent beauty, independent of any applications. Hardy was generally interested in pure mathematics without any concern for solving practical problems.

In his treatise he stated:

> *"I have never done anything "useful". No discovery of mine has made, or is likely to make, directly or indirectly, for good or ill, the least difference to the amenity of the world."*

The pursuit of mathematics can be a challenging mental exercise, not unlike mastering a game of chess, and can be very rewarding when one solves a complex problem, especially in an "elegant" way. An elegant proof in mathematics is one that is generally brief and clever, and does not resort to a lengthy forceful approach.

Hardy was a master in one of the most abstract fields of pure mathematics called *Number Theory*. Its title reveals exactly what it is, the study of the set of positive whole numbers — often referred to as the "natural numbers" — and their myriad relationships. Many of the articles in this book discuss some of those relationships and they can be clearly engaging, such as the fact that there are infinite sets of three whole numbers, a, b and c, that satisfy the Pythagorean Theorem:

---

[3] One unfounded reason is that a woman he was fond of rejected him for a famous mathematician Gosta Mittag-Leffler and Nobel never married. There are also many other speculations as to why there is no Nobel prize in mathematics.

$a^2 + b^2 = c^2$, such as {3,4,5}, {5,12,13}, {8,15,17} etc. See "The Pythagorean Theorem" (page 209). However, there are *no* sets of three whole numbers x, y and z that can be found to satisfy this relationship: $x^n + y^n = z^n$ when n is any number larger than 2. This problem was first proposed by the French Mathematician Pierre de Fermat (1607–1665) in 1637, but no clear proof emerged until more than three centuries later, in 1994, when a proof was provided by the British mathematician Andrew Wiles (1953–). See the previous unit "Fermat's Last Theorem".

A primary way to classify numbers is by divisibility. A number is either prime or composite:

Primes: 2, 3, 5, 7, 11, 13, 17, 19, 23, 29, 31, 37, ...

Composites: 4, 6, 8, 9, 10, 12, 14, 15, 16, 18, 20, ...

Note that the number 1 is neither prime nor composite. Prime numbers have intrigued mathematicians for centuries and they have been studied endlessly. One of the earliest questions to arise in number theory is whether there are an infinite number of primes, and it was proved over 2,000 years ago by Euclid. A simple and elegant indirect proof, similar to Euclid's, is as follows:

Suppose there is a finite number of primes: $p_1, p_2, p_3, \ldots p_n$.

Consider the number which is the product of all the primes: $(p_1)(p_2)(p_3) \cdots (p_n)$.

Add 1 to this number to obtain the number: $(p_1)(p_2)(p_3) \cdots (p_n) + 1$, which is a number that is greater than the product of all the primes and not equal to any of them; so it should not be prime. Therefore, one of the primes above, p_i, $(1 \leq i \leq n)$ should be a factor of this number. However, this cannot be possible, because, while it will divide into the product of all the primes, it will not divide into 1, and there will then be a remainder of $\frac{1}{p_i}$. Hence, this leads to a logical contradiction, which means the original assumption must not be true and there cannot be a finite number of primes. This does not necessarily mean that the number which is the product of all previous primes $(p_1)(p_2)(p_3) \cdots (p_n) + 1$ is prime. The proof only uses the fact that one of the finite primes must divide this number and because it leads to a contradiction the number of primes cannot be finite. At this point, it is interesting to note that if a number $(p_1)(p_2)(p_3) \cdots (p_n) + 1$ is also prime, then it is considered a *primorial prime*, which is a prime number that is equal to the product of all the previous primes ± 1. For example, 5 is a primorial prime because it is equal to $(2 \times 3) - 1$ and 7 is a primorial prime because it is equal to $(2 \times 3) + 1$.

The first eleven primorial primes are:

2, 3, 5, 7, 29, 31, 211, 2309, 2311, 30029, 200560490131, ...

The reason 2 is the first primorial prime is a little elusive. There are no primes preceding 2 so the product of previous primes is considered the empty product and is equal to 1 (not zero), and then $2 = 1 + 1$.

Notice the primorial primes soon become extremely large as they increase. As of March 2018, the largest known primorial prime is the incredible product of the first 85,586 primes −1, and contains an amazing 476,311 digits.

A second basic question about primes concerns the gap between successive primes. For example, the gap between the primes 3 and 5 is 2 while the gap between the primes 7 and 11 is 4. It has been shown that the gap between primes approaches infinity as the numbers increase. See Figure 6.12, which shows how the gaps increase for the first 300 primes. However, there does not appear to be a uniform increase in the spacing of primes and the gaps can vary significantly.

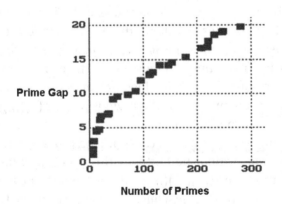

Figure 6.12   Gaps between primes

The first 60 prime gaps are as follows:

1, 2, 2, 4, 2, 4, 2, 4, 6, 2, 6, 4, 2, 4, 6, 6, 2, 6, 4, 2, 6, 4, 6, 8, 4, 2, 4, 2, 4, 14, 4, 6, 2, 10, 2, 6, 6, 4, 6, 6, 2, 10, 2, 4, 2, 12, 12, 4, 2, 4, 6, 2, 10, 6, 6, 6, 2, 6, 4, 2, ...

The only odd prime gap is the first one between 2 and 3 which is 1. All the other prime gaps are even because all the other primes are odd numbers, and therefore their spacing is even. Another observation is that the gap between 3 and 5, and the gap between 5 and 7, are the only pair of consecutive gaps equal to 2.

The average gap between primes does increase as they get larger, but there does not appear to be any clear pattern in the increase in spacing of the prime gaps.

A large amount of study and dedication is usually required to prove a new mathematical concept. Many attempts can be made to prove an idea that appears to be true and yet without proof it can remain a *conjecture* or an open problem (see Goldbach's Conjecture page 343). We close this topic by considering the famous twin prime conjecture, also known as Polignac's conjecture, which asserts that there are infinitely many *twin primes*, which is a pair of primes that differ by 2 gaps, such as (3, 5), (5, 7), (41, 43) etc. Five is the only prime that belongs to two twin primes. A very interesting result is that every twin prime pair greater than (3, 5) can be shown to be of the form $(6n-1, 6n+1)$, where n is any positive whole number or natural number.

As can be seen in the list of prime gaps above, there are many twin primes, but as with primes themselves, twin primes get rarer as the numbers increase. Several mathematicians feel strongly that the twin prime conjecture is a true conjecture. However, it is felt that the proof may be beyond our present capabilities and the twin prime conjecture remains unsolved.

UNIT 91

The Fibonacci Numbers

Perhaps the most ubiquitous numbers in all of mathematics are known as the Fibonacci numbers. They seem to come up in every imaginable aspect of everyday life, such as biology, art, architecture, finance, and obviously almost all branches of mathematics. Let us start at the beginning when they first appeared with the early occurrence of our Hindu-Arabic numerals in Western Europe. It was in a book entitled *Liber abaci*, written in 1202 by Leonardo of Pisa (better known as Fibonacci). He begins the introduction of the book with the following:

> *"... these are the nine figures of the Indians 9, 8, 7, 6, 5, 4, 3, 2, 1. With these nine figures, and with the symbol, 0, which in Arabic is called zephirum, any number can be written, as will be demonstrated below."*

Fibonacci's numerals appear in *Liber Abaci* as shown in Figure 6.13. He refers to the numbers as "Indian" numbers, since apparently the numbers that he saw written right to left in the Arabic World came originally from the Hindus in India.

Figure 6.13   Fibonacci's Numerals

Fascinated as he was by the arithmetic calculations used in the Islamic world, Fibonacci, in this book, first introduced the system of "casting out nines" as a check for arithmetic accuracy.[4] Even today it still comes in handy, although a bit cumbersome. However, the real legacy that Fibonacci achieved resulted from a mathematical problem that he posed in Chapter 12 of his book regarding the regeneration of rabbits. Following a prescribed process for the rabbits' regeneration, namely, that a pair of rabbits requires one month to mature in order to generate another pair of rabbits. Figure 6.14 shows a schematic for this process, where A represents a pair of adult rabbits, which are then able to generate a second pair of baby rabbits (B). The list of pairs of rabbits each month generates the famous Fibonacci numbers.

Month	Pairs	No. of Pairs of Adults (A)	No. of Pairs of Babies (B)	Total Pairs
Jan. 1		1	0	1
Feb. 1		1	1	2
Mar. 1		2	1	3
Apr. 1		3	2	5
May 1		5	3	8
June 1		8	5	13
July 1		13	8	21
Aug. 1		21	13	34
Sept. 1		34	21	55
Oct. 1		55	34	89
Nov. 1		89	55	144
Dec. 1		144	89	233
Jan. 1		233	144	377

Figure 6.14

---

[4] "Casting out nines" refers to an arithmetic check that tells you if your answer is possibly correct. The process requires taking bundles of nine away from the sum or subtracting a specific number of nines from this sum. See Unit 37 "Divisibility by Prime Numbers" for further information on this technique.

When one lists the number of pairs of rabbits each month, the following list results: 1, 1, 2, 3, 5, 8, 13, 21, 34, 55, 89, 144, ... These are the so-called Fibonacci numbers[5], which have countless properties and applications. To generate these numbers, one needs only to add two consecutive numbers to get the next number. For example, $1+1=\mathbf{2}, 1+2=\mathbf{3}, 2+3=\mathbf{5}, 3+5=\mathbf{8}, 5+8=\mathbf{13}, 8+13=\mathbf{21}$, and so on.

There are countless applications and relationships of the Fibonacci numbers. Here are just a few of these unusual properties to merely whet your appetite. We will use the symbol F_n to represent the nth Fibonacci number.

1. The sum of any ten consecutive Fibonacci numbers is divisible by 11:

$$11 \mid (F_n + F_{n+1} + F_{n+2} + \ldots + F_{n+8} + F_{n+9}).$$

2. Consecutive Fibonacci numbers are relatively prime:

    That is, their greatest common divisor is 1.

3. Fibonacci numbers in a composite number position (with the exception of the 4'h Fibonacci number) are also composite numbers. Another way of saying this is that if n is not prime then F_n is not prime (with the condition that n ≠ 4, since $F_4 \neq = 3$, which is a prime number).

4. The sum of the first n Fibonacci numbers is equal to the Fibonacci number that is two numbers further along in the sequence minus one:

$$\sum_{i=1}^{n} F_i = F_1 + F_2 + F_3 + F_4 + \cdots + F_n = Fn+2-1.$$

5. The sum of the consecutive even-positioned Fibonacci numbers is 1 less than the Fibonacci number that follows the last even number in the sum:

---

[5] For more about the Fibonacci Numbers see *The Fabulous Fibonacci Numbers*, A. S. Posamentier and I. Lehmann, Amherst, NY: Prometheus Books, 2007.

$$\sum_{i=1}^{n} F_{2i} = F_2 + F_4 + F_6 + \cdots + F_{2n-2} + F_{2n} = F_{2n+1} - 1.$$

6. *The sum of the consecutive odd-positioned Fibonacci numbers is equal to the Fibonacci number that follows the last odd number in the sum:*

$$\sum_{i=1}^{n} F_{2i-1} = F_1 + F_3 + F_5 + \cdots + F_{2n-3} + F_{2n-1} = F_{2n}.$$

7. *The sum of the squares of the Fibonacci numbers is equal to the product of the Fibonacci numbers of the last number and the next number in the sequence:*

$$\sum_{i=1}^{n} (F_i)^2 = F_n F_{n+1}.$$

8. *The difference of the squares of two alternate[6] Fibonacci numbers is equal to the Fibonacci number in the sequence whose position number is the sum of their position numbers:*

$$F_n^2 - F_{n-2}^2 = F_{2n-2}.$$

9. *The sum of the squares of two consecutive Fibonacci numbers is equal to the Fibonacci number in the sequence whose position number is the sum of their position numbers:*

$$F_n^2 + F_{n+1}^2 = F_{2n+1}.$$

10. *For any group of four consecutive Fibonacci numbers the difference of the squares of the middle two numbers is equal to the product of the outer two numbers.* Symbolically, we can write this as
$$F_{n+1}^2 - F_n^2 = F_{n-1} \cdot F_{n+2}$$

11. *The product of two alternating Fibonacci numbers is 1 more or less than the square of the Fibonacci number between them* ± 1:

$$F_{n-1} F_{n+1} = F_n^2 + (-1)^n$$

(*If n is even, the product is 1 more; if n is odd the product is 1 less.*)

---

[6] A Fibonacci number with number n and the Fibonacci number two before it.

12. *The difference between the square of the selected Fibonacci number and the various products of Fibonacci numbers equidistant from the selected Fibonacci number is the square of another Fibonacci number:*

$$F_{n-k}F_{n+k} - F_n^2 = \pm F_k^2, \text{ where } n \geq 1, \text{ and } k \geq 1$$

As was mentioned earlier, the Fibonacci numbers can be found in a wide variety of fields. As an example, the Fibonacci numbers can be seen in many aspects of biology or nature. For instance, the Fibonacci numbers can be seen in most spirals that occur in nature, such as the spirals formed by the bracts of a pineapple, where two of the spirals have 8 spirals and 13 spirals in one direction and 5 spirals in the other direction, as shown in Figure 6.15. These are three consecutive Fibonacci numbers.

Figure 6.15

Furthermore, the famous Golden Ratio (see page 223) can also be generated by the Fibonacci numbers. This can be done by continually taking the quotient of consecutive Fibonacci numbers in either direction. One gets the Golden Ratio, $\frac{1+\sqrt{5}}{2} = 1.61803398874989\ldots$, in one direction and its reciprocal $\left(\frac{1-\sqrt{5}}{2} = 0.61803398874989\ldots\right)$ in the other direction as shown in Figure 6.16.

The Ratios of Consecutive Fibonacci Numbers [33]

$\dfrac{F_{n+1}}{F_n}$	$\dfrac{F_n}{F_{n+1}}$
$\dfrac{1}{1} = 1.000000000$	$\dfrac{1}{1} = 1.000000000$
$\dfrac{2}{1} = 2.000000000$	$\dfrac{1}{2} = 0.500000000$
$\dfrac{3}{2} = 1.500000000$	$\dfrac{2}{3} = 0.666666667$
$\dfrac{5}{3} = 1.500000000$	$\dfrac{3}{5} = 0.600000000$
$\dfrac{8}{5} = 1.600000000$	$\dfrac{5}{8} = 0.625000000$
$\dfrac{13}{8} = 1.625000000$	$\dfrac{8}{13} = 0.615384615$
$\dfrac{21}{13} = 1.615384615$	$\dfrac{13}{21} = 0.619047619$
$\dfrac{34}{21} = 1.619047619$	$\dfrac{21}{34} = 0.617647059$
$\dfrac{55}{34} = 1.617647059$	$\dfrac{34}{55} = 0.618181818$
$\dfrac{89}{55} = 1.618181818$	$\dfrac{55}{89} = 0.617977528$
$\dfrac{144}{89} = 1.617977528$	$\dfrac{89}{144} = 0.618055556$
$\dfrac{233}{144} = 1.618055556$	$\dfrac{144}{233} = 0.618025751$
$\dfrac{377}{233} = 1.618025751$	$\dfrac{233}{377} = 0.618037135$
$\dfrac{610}{377} = 1.618037135$	$\dfrac{377}{610} = 0.618032787$
$\dfrac{987}{610} = 1.618032787$	$\dfrac{610}{987} = 0.618034448$

Figure 6.16

We have merely provided here a brief introduction to one of the most popular topics in mathematics — the Fibonacci Numbers. Since 1963, the Fibonacci Association publishes the Fibonacci Quarterly[7] covering a wide variety of applications of the Fibonacci numbers in mathematics and beyond.

UNIT 92

Diophantine Equations

Typically, in an elementary algebra course when an equation is given with 2 variables, say x and y, it is expected there will be a second equation with the same 2 variables so that the two equations can be solved simultaneously. What is sought after is a pair of values for these 2 variables to satisfy both equations. However, what seems to have been omitted from the school curriculum is to show how one can "solve" an equation with 2 variables, where no second equation is provided. Although there is very little known about the ancient Greek mathematician Diophantus of Alexandria whose approximate dates of life are 207–291 CE. We attribute to him the methods of solving such equations with integral solutions. These attributions are largely the result of a series of books that he wrote called Arithmetica. Of the original 13 books in this series, only 6 have survived to the present date. Diophantus' work has been honored through the ages and provides a method of solving certain types of equations as we shall see in the following.

Perhaps the best way to understand this kind of equation is to consider a problem that generates a Diophantine equation: *How many combinations of 6-cent and 8-cent stamps can be purchased for 5 dollars?*

As we begin, we realize that there are 2 variables that must be determined, say x and y. Letting x represent the number of 8-cent stamps and y represent the number of 6-cent stamps, the equation: $8x + 6y = 500$ should follow. This can then be reduced to $4x + 3y = 250$.

---

[7] See www.fq.math.ca.

At this juncture we realize that although this equation has an infinite number of solutions, it may or may not have an infinite number of *integral* solutions; moreover, it may or may not have an unlimited number of *positive integral* solutions (as called for by the original problem). The first problem to consider is whether integral solutions, in fact, actually exist.

In order to determine if an integral number of solutions actually exist, a useful theorem may be employed. It states that if the greatest common factor of a and b is also a factor of k, where a, b, and k are integers, then there exist an infinite number of integral solutions for x and y in $ax + by = k$. Equations of this type, whose solutions must be integers are known as *Diophantine equations* in honor of the Greek mathematician Diophantus (ca. 201–285 CE), who wrote about them in a series of books entitled *Arithmetica*.

Since the greatest common factor of 3 and 4 is 1, which is a factor of 250, there exist an infinite number of integral solutions to the equation $4x + 3y = 250$. The question which we must now consider is how many (if any) *positive* integral solutions exist for this equation? One possible method of solution is often referred to as Euler's method, named after the Swiss mathematician Leonhard Euler (1707–1783). To begin, we should solve for the variable with the coefficient of least absolute value; in this case, y, which gives us $y = \frac{250-4x}{3}$. This can be rewritten to separate the integral parts as $y = 83 + \frac{1}{3} - x - \frac{x}{3} = 83 - x + \frac{1-x}{3}$. We now introduce another variable, t, and let $t = \frac{1-x}{3}$. Solving for x, yields $x = 1 - 3t$. Since there is no fractional coefficient in this equation, the process does *not* have to be repeated (i.e. each time introducing new variables, as with t, above). Now substituting for x in the above equation yields $y = \frac{250-4(1-3t)}{3} = 82 + 4t$. For various integral values of t, corresponding values for x and y can be generated. A table of values such as that shown in Figure 6.17 might prove useful.

t	...	−2	−1	0	1	2	...
x	...	7	4	1	−2	−5	...
y	...	74	78	82	86	90	...

Figure 6.17

Perhaps by generating a more extensive table, we would notice for which values of t would allow us to obtain positive integral values for x and y. However, such a procedure for determining the number of positive integral values of x and y is not very elegant. Therefore, we will solve the following inequalities since there are positive values. So, we then have $1 - 3t > 0$ and then $t < \frac{1}{3}$. For the other inequality $82 + 4t > 0$, and then $t > -20\frac{1}{2}$. This can then be combined and stated as $-20\frac{1}{2} < t < \frac{1}{3}$, and indicates that there are 21 possible combinations of 6-cent and 8-cent stamps that can be purchased for 5 dollars. (N. B. Values that t can take on are $0, -1, -2, -3, \ldots, -19, -20$.)

To get a better grasp on this neglected topic, we offer here another Diophantine equation along with its solution. This should serve to further support this important algebraic aspect. We shall consider solving the Diophantine equation $5x - 8y = 39$.

First, we solve for x, since its coefficient has the lower absolute value of the two coefficients.

$$x = \frac{y+39}{5} = y + 7 + \frac{3y+4}{5}$$

Let $t = \frac{3y+4}{5}$, then we solve for y:

$$y = \frac{5t-4}{3} = t - 1 + \frac{2t-1}{3}$$

Since we still have a fraction in the last equation, let $u = \frac{2t-1}{3}$, and then we solve for t:

$$t = \frac{3u+1}{2} = u + \frac{u+1}{2}.$$

Once again, since there is a fraction in the last equations, we let $v = \frac{u+1}{2}$, and then solve for u:

$u = 2v - 1$. We may now reverse the process because the coefficient of v is an integer. Therefore, substituting in reverse order, we get: $t = \frac{3u+1}{2}$. Thus, $t = \frac{3(2v+1)+1}{2} = 3v - 1$.

v	...	-2	-1	0	1	2	...
x	...	-13	-5	3	11	19	...
y	...	-13	-8	-3	2	7	...

Figure 6.18

Furthermore, $y = \frac{5t-4}{3}$. Therefore, $y = \frac{5(3v-1)-4}{3} = 5v - 3$.
Since $x = \frac{y+39}{5}$, we get $x = \frac{8(5v-3)+39}{5} = 8v + 3$.

Now, with our values of x and y: $x = 8v + 3$, and $y = 5v - 3$, we have solved the equation and can then set up a table of values as shown in Figure 6.18.

We notice that since we were not restricted to positive values of x and y, as we were in the previous example, we have many solutions as shown in Figure 6.18. This is an important aspect of algebra, which dates back to ancient Greek times. It is unfortunate that this topic seems to have been left out of the school curriculum.

UNIT 93

The Goldbach Conjecture

Often times conjectures are made in mathematics that lead to new discoveries once they are proven to be correct. On page 136, we show a conjecture by the French mathematician Alphonse de Polignac (1817–1890) that turned out not to be correct because a counter example was found. There are, however, conjectures which have not been proven or disproven to be true. Perhaps the most famous conjecture was made by the Prussian mathematician Christian Goldbach (1690–1764) in a 1742 letter to the famous Swiss mathematician, Leonhard Euler, where he posed the following relationship, which to this day has yet to be solved. It is known as *Goldbach's Conjecture* and states the following:

Every even number greater than 2 can be expressed as the sum of two prime numbers.

Even numbers greater than 2	Sum of two prime numbers
4	2 + 2
6	3 + 3
8	3 + 5
10	3 + 7
12	5 + 7
14	7 + 7
16	5 + 11
18	7 + 11
20	7 + 13
48	19 + 29
100	3 + 97

Figure 6.19

Figure 6.19 shows a list of some even numbers and their corresponding sum of two prime numbers.

Mathematicians have tried for centuries to prove that this is true for all even numbers, but no one has yet been successful. At the same time, no one has found a counterexample that would indicate that this relationship is not true for all numbers. Hence the conundrum!

Goldbach later produced a follow-up conjecture, commonly referred to as Goldbach's Second Conjecture, which is also never been proven or disproven. This conjecture is as follows:

Every odd number greater than 5 is the sum of three primes.

Let us consider the first few odd numbers (Figure 6.20):

You may wish to see if there is a pattern here and generate other examples. These unsolved problems have tantalized many mathematicians over the centuries and although no solution has yet been found, there is more evidence (with the help of computers) that these must be true since no counterexample has been found. Interestingly, the efforts to solve these have led to some important discoveries in mathematics that might have gone hidden without this impetus. For those of us who enjoy mathematics, they are intriguing and a source of curiosity.

Odd numbers greater than 5	Sum of three prime numbers
7	2 + 2 + 3
9	3 + 3 + 3
11	3 + 3 + 5
13	3 + 5 + 5
15	5 + 5 + 5
17	5 + 5 + 7
19	5 + 7 + 7
21	7 + 7 + 7
51	3 + 17 + 31
77	5 + 5 + 67
101	5 + 7 + 89

Figure 6.20

UNIT 94

Triskaidekaphobia

The number 13 is usually associated with being an unlucky number. Buildings with more than thirteen stories will typically omit the number 13 from the floor numbering. This is immediately noticeable in an elevator, where there is no button for 13. You can certainly think of other examples where the number 13 is associated with bad luck such as when the 13th of a month turns up on a Friday, then it is considered a particularly a bad omen. This may derive from the belief that there were *thirteen* people present at the Last Supper, which resulted in the Jesus' crucifixion on a *Friday*. Do you think that the 13th comes up on a Friday with equal regularity as on the other days of the week? You may be astonished that, lo and behold, the 13th comes up a bit more frequently on Friday than on any other day of the week.

This fact about Friday the 13th was first published by B.H. Brown,[8] where he stated that the Gregorian calendar follows a pattern of leap

---

[8] "Solution to Problem E36." *American Mathematical Monthly*, 1933, vol. 40, p. 607.

years, repeating every 400 years. The number of days in one four-year cycle is $3 \times 365 + 366$. Therefore, in 400 years there are

$100 \times (3 \times 365 + 366) - 3 = 146{,}097$ days. (Note that century years that are divisible by 400 are not leap-years, hence the deduction of 3.) This total number of days is exactly divisible by 7. Since there are 4800 months in this 400-year cycle, the 13th comes up 4800 times according to the following table. Interestingly enough, the 13th comes up on a Friday more often than on any other day of the week. Figure 6.21 summarizes the frequency of the 13th appearing on the days of the week.

Then there are well-known triskaidekaphobics[9], such as Napoleon Bonaparte, Herbert Hoover, Mark Twain, Richard Wagner, and Franklin Delano Roosevelt. It was said that Roosevelt would not invite 13 people around the dinner table and would not begin any trip on the 13th of the month. Richard Wagner (1813–1883), the famous German composer who revolutionized music, might have attributed his phobia for the number 13 to the Norse legends where 12 gods were gathered in Valhalla and when the 13th troublesome god, Loge, arrived and caused deadly mischief, the number 13 was seen as unlucky. To what extent Wagner was influenced by this legend will remain unknown; however, for him the number 13 seemed to have played a curious role. Wagner included Loge in his famous *Der Ring des Nibelungen*, although not as a bad person, but merely as a manipulator. To understand how the number 13 was manifested in Wagner's life consider the following:

Wagner was born in 18**13**, which has a digit sum of **13**.
Wagner died on February **13**, 1883, which was the **13**[th] year of the unification of Germany.
Wagner wrote **13** operas during his lifetime.
Richard Wagner's name consists of **13** letters.
Wagner's opera, Tannhäuser, was completed on April **13**, 1845.

---

[9] *A phobia for the number 13.*

Day of the week	Number of 13s	Percent
Sunday	687	14.313
Monday	685	14.271
Tuesday	685	14.271
Wednesday	687	14.313
Thursday	684	14.250
Friday	*688*	*14.333*
Saturday	684	14.250

Figure 6.21

On May **13**, 1861 Wagner premiered Tannhäuser in Paris during his **13**-year exile from Germany.

The grand opening of Wagner's festival opera house in Bayreuth, Germany was opened on August **13**, 1876.

Wagner completed his last opera, Parsifal, on January **13**, 1882.

Wagner's Last day in Bayreuth, the city in which he built his famous festival opera house, was September **13**, 1882.

On a more positive note, Ludwig van Beethoven's Pathetique Sonata, is Opus No. 13. We also know that there are 13 cards in each suit of a deck of playing cards. At age 13, in the Jewish religion, a boy becomes a man at a bar mitzvah, clearly a positive acceptance of the number 13.

The original flag of the United States of America had 13 stripes and 13 stars to commemorate the initial 13 colonies that formed the United States. Today's flag has retained the 13 stripes but now has 50 stars representing the number states.

However, on the brighter side, in mathematics, the number 13 is sometimes referred to as a "star number," since 13 dots can be arranged in the shape of a hexagram or 6-corner star, as we can see in Figure 6.22.

In mathematics, there are also some curious specialties such as the pattern we see here: $13^2 = 169$ and $31^2 = 961$. It is also curious that the prime number 13 is one of those numbers that when it is reversed is also a prime number, namely 31. The relationship that exists between the number 13 and its reversal 31, is that 13 is the

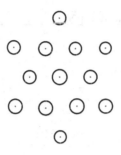

Figure 6.22

smallest number that can be expressed as the sum of 2 prime numbers squared, namely, $2^2 + 3^2 = 13$.

The number 13 is one of 3 Wilson primes, where the other two such primes are the numbers 5 and 563. A Wilson prime, named for the British mathematician John Wilson (1741–1793), is a prime number n, which is a factor of $(n - 1)! + 1$. In this case, when $n = 13$, we have $(13 - 1)! + 1 = 479,001,600 + 1 = 479,001,601 = 13 \times 36,846,277$. We can even take this a step further by noticing that $(13 - 1)! + 1$ also has 13^2 as a factor. Namely $479,001,601 = 13^2 \times 2,834,329$.

While we are considering 13^2, we can notice that it is the difference of 2 consecutive cubes as follows: $13^2 = 169 = 8^3 - 7^3$. Despite its negativity in cultural contexts, the number 13 still has a few redeeming values from a mathematical standpoint.

Powers related to 13 can lead to some interesting relationships. For example:

$13^{13} - 13 + 1 = 3,028,751,046,592,241$ is a prime number.
13^2 is a factor of $12! + 1 = 479,001,601 = 13^2 \times 2,834,329$.
$13^2 = 7 + 8 + 9 + 10 + 11 + 12 + 13 + 14 + 15 + 16 + 17 + 18 + 19$.
$13^2 = + 8^3 + 7^3$.
$2^{13} - 13 = 8179$, which is a prime number.
$13 \times 2 = -1^3 + 3^3 = 27 - 1 = 26$.

We can use the first 3 prime numbers to represent 13 as: $2^3 + 5 = 13$.

The sum of the first 6 prime numbers: $2 + 3 + 5 + 7 + 11 + 13 = 41$, is the 13th prime number.

If we divide 13 by each of the first 13 prime numbers, the sum of the remainders $1 + 1 + 3 + 6 + 2 = 13$.

This number 13 is the smallest prime number where the sum of his digits is a perfect square, namely 4.

The sum of the first 13 prime numbers $2 + 3 + 5 + 7 + 11 + 13 + 17 + 19 + 23 + 29 + 31 + 37 + 41 = 238$, whose digit sum is 13.

A prime number can be generated by applying the number 13 in a rather curious way. The numerator of the following fraction has 14 digits of which 13 are 3s: $\frac{13333333333333}{13} = 1{,}025{,}641{,}025{,}641$, which is a prime number.

There are exactly 13 prime numbers consisting of 4 consecutive digits, and they are: 1423, 2143, 2341, 2543, 4231, 4253, 4523, 4567, 4657, 5647, 5867, 6547, 6857.

The number of composite numbers between 13 and its reversal, 31, is 13. They are: 4, 6, 8, 9, 10, 12, 14, 15, 16, 18, 20, 21, 22, 24, 25, 26, 27, 28.

We can represent the number 13 in terms of just prime numbers is as follows:

$(5 \times 11) - (2 \times 3 \times 7)$. A motivated reader might like to try to create the number 13 in other ways using only prime numbers.

The number 13 is the only prime number where its double, 26, increased by 1 is a perfect cube; $2 \times 13 + 1 = 27 = 3^3$.

A nifty palindrome can be created which begins and ends with 13.

131,211,109,876,543,212,345,678,910,111,213. What makes this palindrome special is that it is a prime number.

In geometry, there is a unique right triangle whose longest side has length 13, and where the area is numerically equal to the perimeter. The sides of this triangle are 5, 12 and 13. The area of this triangle is $\frac{1}{2} \times 5 \times 12 = 30$, while the perimeter is equal to $5 + 12 + 13 = 30$.

UNIT 95

The Parabola — A Very Special Curve

One of the first curves, other than the straight line, that one encounters in high school geometry is the *parabola*. Though its equation is a simple quadratic expression: $y = ax^2 + bx + c$, it possesses an array of applications, some of which many people often experience in their daily lives without being aware of it. Let's examine this exceptional curve which is shown in Figure 6.23 and reveal its features and its versatility. We define such a curve by how the points are located. For example, the circle is defined as a curve which contains all the points equidistant from a fixed point, called the center. In a similar way, the parabola is defined as a curve which contains all the points equidistant from a fixed point called the *focus* and a line called the *directrix* (see Figure 6.23).

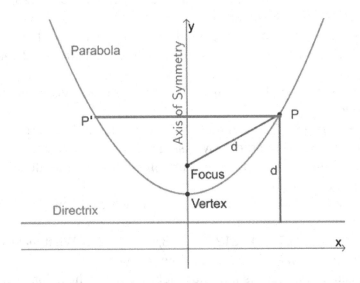

Figure 6.23   The Parabola

In the figure, the distance *d* of point *P* to the focus is equal to the distance *d* to the *directrix*. The turning point of the parabola is called the *vertex*. As one travels further away from the vertex, the distances to the focus and the directrix keep increasing so the curve continues

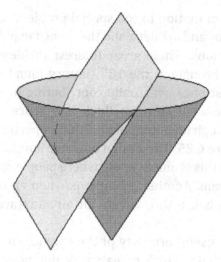

Figure 6.24   Parabola as a Conic section

to widen indefinitely. A vertical line (perpendicular to the x axis) passing through the vertex is called the *axis of symmetry* because it cuts the curve into two symmetrical parts. For every point P on one side of the *axis of symmetry*, there is a symmetrical point P' directly on the opposite side. The parabola is also considered a *conic section* because when a cone is cut by a plane parallel to its side the intersection is a parabola as shown in Figure 6.24. The name *parabola* may have gotten its name from this property. In Greek *para* can mean "alongside of" and *bola* "throwing", so parabola literally means "beside-throwing" as in throwing two objects together — in this case the plane and the cone.

The first very important application of the parabola has to do with the laws of gravity.

Any object moving freely through the earth's gravitational field will travel in a path which is the arc of a parabola.

The only rare exception is if its motion is directly up or down, in which case its path is a straight vertical line. With the advancement of mathematics in the 18th century, this property of the parabola contributed to major military and political repercussions. Armies applied the mathematics of motion to the special problems of the range and trajectories of guns and artillery, and the French in particular, excelled in these applications. Their great interest in ideas and advances in mathematics throughout the 18th century found its way into the field of military science eventually contributing to the successes of Napoleon's armies. Missiles, satellites and space exploration rely heavily on a thorough understanding of the properties of the parabola as shown in Figure 6.25. The design of water fountains utilizes parabolic motion, and this is one example where people can see a parabola in the water stream. Another example is when an object, such as a basketball or baseball, is thrown into the air and travels in the path of a parabola.

Another very useful property of the parabola has to do with the focus of the parabola, which reveals why this point has that name. When a light ray reflects off a surface, such as a mirror, the angle the ray makes with the perpendicular to the surface, called the angle of incidence, equals the angle of reflection. The shape of a parabolic

Figure 6.25 Parabolic Rocket Trajectory

reflector is such that any light ray, or electromagnetic ray, incident on the surface and parallel to the axis, will be reflected to the focus, as shown in Figure 6.26. The reverse is also true, that any ray emanating from the focus will be reflected parallel to the axis.

This leads to a myriad of applications for a surface generated by revolving a parabola about its axis, which is called a circular paraboloid. Some applications of which are shown in Figure 6.27 where the parabolic dishes reflect all the parallel rays to the focus thereby amplifying the signal.

The ability of a parabolic reflector to do the reverse and project a parallel beam of light from a source at its focus finds significant application in flashlights, searchlights and car headlights. Parabolic reflector antennas are also used to direct a narrow beam of radio waves to satellite dishes and microwave relay stations which can help to locate aircraft, ships and vehicles. Parabolic microphones can pick up distant sounds and eavesdrop on private conversations for legal and government use.

In early Greece, the mathematician Diocles (240 BCE–180 BCE) explained the principle of a parabolic reflector in his book *On Burning Mirrors* and proved that it focused a parallel beam of light to a point. In the 17th century, Isaac Newton invented the reflecting telescope which used a parabolic mirror to focus images of stellar objects (Figure 6.28). It proved superior to the refracting telescopes of the

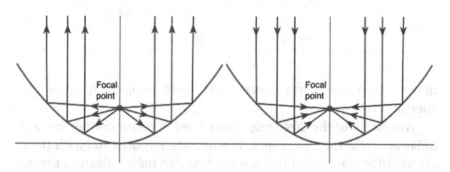

Figure 6.26   Parabolic reflection

One of the world's largest solar parabolic dishes at the Ben Gurion Solar Energy Center in Israel.

One of the largest radio telescopes in Parkes, Australia

Figure 6.27

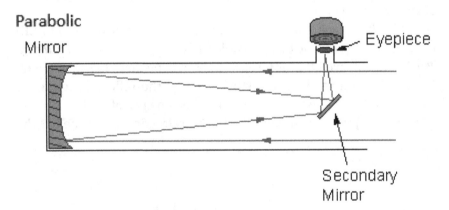

Figure 6.28   Reflecting telescope

times which used only lenses and were subject to chromatic aberration.

Almost all of the large telescopes used in astronomical research today are reflecting telescopes. The German Physicist Heinrich Hertz (1857–1894) fabricated the world's first parabolic reflector antenna which today has become ubiquitous in its use to focus light, sound, solar energy and radio waves. So, the next time you see a TV satellite dish, or use a water fountain, or see a ball thrown into the air, carefully focus on the parabolic arc that is generated.

UNIT 96

Kepler's Three Famous Laws

Johannes Kepler (1571–1630) (see Figure 6.29) was a German mathematician and astronomer who excelled as a geometer. Through laborious geometric constructions he arrived at three famous laws of planetary motion:

1. **Law of Ellipses**: The path of a planet is an ellipse where the Sun is at one focus.

2. **Law of Equal Areas**: The area swept out by a radius vector drawn from the Sun sweeps out equal areas in equal intervals of time in the planet's orbit.

3. **Law of Harmonies**: The ratio of the square of the time it takes for a planet to complete its orbit, to the cube of its average distance from the sun, is the same for each planet.

Figure 6.29   Johannes Kepler

The original theory of the solar system was formulated by the Egyptian mathematician Ptolemy (100 CE–170 CE) as a geocentric system where the planets and the sun revolved around the Earth. This religious motivated belief remained in force for almost 1500 years before it was disputed by the Polish mathematician Nicolaus Copernicus (1473–1543) who proposed a heliocentric (Sun-centered) system, which naturally took time to become accepted by the world. Kepler was greatly influenced by Copernicus and the Danish astronomer Tycho Brahe (1546–1601) and published his first two laws in 1609 and his third law in 1618, which helped reinforce the heliocentric theory. Tycho Brahe was the last of the major naked-eye astronomers working without telescopes but had an incredible power of observation. Though Kepler drew upon Brahe's data, he realized that empirical observations would not be sufficient for the theoretical structure he needed, and therefore, relied principally on a geometrical analysis.

Kepler's First Law

An ellipse is a curve where the sum of the distances from any point on the curve to two fixed points, called the foci, is the same for any point on the curve. Figure 6.30 shows the elliptical orbit of a planet with the Sun at one focus and the sum of the distances, $d_1 + d_2$, from two randomly selected points on the curve being equal to the same constant, which is the major axis of the ellipse.

The *Law of Ellipses* completely contradicted Ptolemy's orbits, which were complicated curves with attached *epicycles.* The epicycles were loops with backward motion that accounted for the apparent retrograde motion of the planets that is experienced because the speed of the Earth's orbit is different from the orbital speed of any other planet.

Kepler's Second Law

Kepler's *Law of Equal Areas* was a brilliant result that clearly explained variations in the motion of the planets. The areal velocity of a planet, which is the area swept out divided by the time, is constant throughout its orbit. Figure 6.31 shows how the speed of a planet becomes

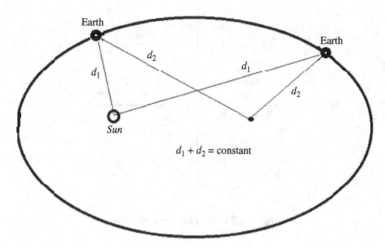

Figure 6.30   Kepler's first law

greater when it is closer to the Sun. This is because it sweeps out less area for the same distance that it travels when it is further away from the Sun. Similarly, the speed slows down as it is further away from the Sun because it sweeps out more area for the same distance that it travels when it is closer to the Sun. In Figure 6.31, the three areas are all equal, $A_1 = A_2 = A_3$, but the distances the planet travels are greater when closer to the Sun and vice versa.

Kepler's Third Law

The first and second laws describe the nature of a single planet's path, whereas the third law, *Law of Harmonies,* compares the motion characteristics of all the planets. It is a remarkable mathematical relationship between the time any planet takes to complete its orbit and its average or mean distance from the Sun. If T is the time for a planet to complete one orbit, called its period, and d is its average distance from the Sun, then the following ratio is the same for each planet $\frac{T^2}{d^3} = constant.$

We illustrate this relationship in the table (Figure 6.32), which gives the orbital periods and average distances from the Sun for the Earth and Mars.

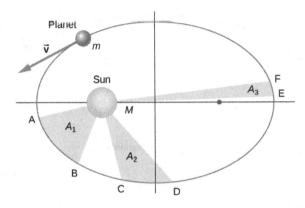

Figure 6.31   Kepler's second law

Planet	Period (T)	Average Distance (d)	$\dfrac{T_2}{d_3}$
Earth	3.156×10^7 s	1.4957×10^{11}	2.977×10^{-19}
Mars	5.93×10^7 s	2.278×10^{11}	2.975×10^{-19}

Figure 6.32

The value for $\frac{T^2}{d^3}$ is remarkably not just the same for all the planets, but also for the moons of the planets and any artificial satellites. Johannes Kepler's genius had a strong influence on Isaac Newton's theory of gravity and the discovery of Calculus.

UNIT 97

Beyond the Quadratic Equation

In elementary algebra, there are two basic formulas that one remembers long after much of the mathematics taught in school fades. The first and most memorable, is the famous Pythagorean Theorem: $a^2 + b^2 = c^2$. Rarely does a high school graduate forget this theorem. The second, and somewhat less memorable, is the quadratic formula which is used to solve the quadratic equation $ax^2 + bx + c = 0$, when one knows the values of the coefficients a, b and c:

$$x = \frac{-b \pm \sqrt{b^2 - 4ac}}{2a}$$

This simple formula is quite easy to apply and beginning algebra students rarely have difficulty with it. One clear property of the formula is that every quadratic equation has two roots, even if the radical is zero and the roots are the same, we say mathematically that it is a double root. Have you ever wondered what the analogous formula is for a general cubic, or third-degree equation, which has four coefficients and looks like this: $ax^3 + bx^2 + cx + d = 0$? If you happen to have taken more than three years of high school mathematics you may have solved certain cubic equations, but the chances are you never saw the general cubic formula, and there is a good reason for it. At first glance, one may think it is probably just a little more complex than the quadratic formula. Well, the simple increase of the exponent from 2 to 3 opens up a much larger "can of worms," so to speak. The three-dimensional world is far more complex than the two-dimensional one. Let us explore the cubic equation and solution, which will help you to appreciate why this is so. The simplest third-degree equation is one that has only the first and last terms $ax^3 + d = 0$. Its solution is easily obtained by solving for x^3 and taking the cube root of each side:

$$x = \sqrt[3]{\frac{d}{a}}$$

However, there are three solutions to every cubic equation, and in this case two of the solutions are complex numbers and contain the imaginary quantity $i = \sqrt{-1}$. For example, if $x^3 = 8$, the three solutions are: $x_1 = 2, x_2 = -1 - i\sqrt{3}, x_3 = -1 + i\sqrt{3}$

As we consider more terms in the cubic equation, the solutions increase in complexity. When we arrive at the general solution in terms of the four coefficients a, b, c and d, it completely overshadows the comparatively simple quadratic formula as you can see in Figure 6.33 where x_1, x_2 and x_3 are the roots.

Clearly, adding one more dimension takes us into a much more complicated world of radicals, cube roots and imaginary numbers. When we move on to a quartic equation, which is of degree four: $ax^4 + bx^3 + cx^2 + dx + e = 0$, you can appreciate that the complexity will increase exponentially and the general solution of a quartic equation, though it exists, doesn't need to be shown and can be left for the reader to research if so inclined.

When mathematicians tackled the general fifth-degree equation, a startling discovery emerged that further demonstrated how convoluted another dimension can be. The solution to the general cubic equation emerged during the Renaissance and is due to the contributions of several mathematicians. Lodovico Ferrari (1522–1565) is

$$x_1 = -\frac{b}{3a}$$
$$-\frac{1}{3a}\sqrt[3]{\frac{1}{2}\left[2b^3 - 9abc + 27a^2d + \sqrt{(2b^3 - 9abc + 27a^2d)^2 - 4(b^2 - 3ac)^3}\right]}$$
$$-\frac{1}{3a}\sqrt[3]{\frac{1}{2}\left[2b^3 - 9abc + 27a^2d - \sqrt{(2b^3 - 9abc + 27a^2d)^2 - 4(b^2 - 3ac)^3}\right]}$$

$$x_2 = -\frac{b}{3a}$$
$$+\frac{1+i\sqrt{3}}{6a}\sqrt[3]{\frac{1}{2}\left[2b^3 - 9abc + 27a^2d + \sqrt{(2b^3 - 9abc + 27a^2d)^2 - 4(b^2 - 3ac)^3}\right]}$$
$$+\frac{1-i\sqrt{3}}{6a}\sqrt[3]{\frac{1}{2}\left[2b^3 - 9abc + 27a^2d - \sqrt{(2b^3 - 9abc + 27a^2d)^2 - 4(b^2 - 3ac)^3}\right]}$$

$$x_3 = -\frac{b}{3a}$$
$$+\frac{1-i\sqrt{3}}{6a}\sqrt[3]{\frac{1}{2}\left[2b^3 - 9abc + 27a^2d + \sqrt{(2b^3 - 9abc + 27a^2d)^2 - 4(b^2 - 3ac)^3}\right]}$$
$$+\frac{1+i\sqrt{3}}{6a}\sqrt[3]{\frac{1}{2}\left[2b^3 - 9abc + 27a^2d - \sqrt{(2b^3 - 9abc + 27a^2d)^2 - 4(b^2 - 3ac)^3}\right]}$$

Figure 6.33   Cubic formula

credited with the discovery of the solution of the quartic equation in 1540, but since his solution required the solution of a cubic to be found, it couldn't be published immediately, and was published together with that of the cubic by Ferrari's mentor, Gerolamo Cardano (1501–1576), in the book *Ars Magna* (1545).

The Norwegian mathematician Neils Abel (1802–1829) worked on quintic equations and in 1824, he came up with the incredible proof that it was impossible to produce a general solution to the quintic equation in terms of radicals. Around the same time, the French mathematician Évariste Galois (1811–1832) developed the theory of Groups, which further elucidated the impossibility of finding radical solutions to *any* equations of order higher than degree four (see Évariste Galois (page 361). Abel, unfortunately, succumbed to ill health at an early age, while Galois died in a duel during a turbulent political time in France. Both these brilliant men unfortunately died in their twenties, a time in people's lives when they tend to make great contributions to science and mathematics. The general cubic and quartic formulas are not actually used to solve higher degree equations as there are far more efficient methods that can be employed. Often a sophisticated trial and error approximation approach, which zeroes in on the solutions, is faster and more direct.

UNIT 98

Évariste Galois — A Short Brilliant Life

History shows us that genius tends to manifest itself at an early age. Albert Einstein, Isaac Newton, Wolfgang Amadeus Mozart and Pablo Picasso all made seminal contributions to their fields in their twenties and earlier. Groundbreaking work in mathematics also occurs at a young age. Perhaps the young mind is most open to an extraordinary ability to apply creativity and imaginative thinking to a problem and is not encumbered by the pressures of past ideas and social norms. One such outstanding example of young genius is that exhibited by the French mathematician and political activist Évariste Galois (1811–1832) (see Figure 6.34). While still in his teens, he created the foundations for a major branch of abstract algebra called *group theory*. In mathematics, a *group* is a set of numbers or elements which has an operation defined for any two of its elements that satisfies just three conditions. The first is that the operation is *associative*. For example, the integers, with the operation of addition, form a group. The *associative operation* for three integers, *a*, *b*, and *c* is the following: $(a + b) + c = a + (b + c)$. In other words, the result is the same if you add the first two integers and then the third, or if you add the last two integers and then the first, keeping the same order. The second condition is that the group has an *identity element*. When an element is operated on by the identity element, the result is the original element. The number zero is the identity element for the addition of integers since: $a + 0 = 0 + a = 0$. The third condition is that each element in the group has an *inverse element*. When an element is operated on by its inverse, the result is the identity element. The inverse element for any integer is the negative of the integer: $a + (-a) = 0$.

On the other hand, the set of integers with the operation of multiplication is not a group. While the operation of multiplication is

Figure 6.34   Évariste Galois at about 15 years of age

associative, and the number one is an identity element, there does not exist an inverse for every number. The inverse of any integer a is equal to $\frac{1}{a}$ since $(a)(\frac{1}{a}) = 1$, but $\frac{1}{a}$ is not an integer and is therefore not in the group. These three simple conditions give rise to many complex mathematical structures and form a basis for modern abstract algebra.

Évariste Galois lived at a politically tumultuous time in France when a coup d'état was staged against the majority liberal party touching off the Revolution of 1830. Galois' political activity resulted in several stays in prison where he formulated some of his mathematics involving the theory of equations. Upon his release from his last stay in prison on April 29, 1832, he somehow found himself challenged into a gun duel. The duel was possibly due to his affair with Stéphanie-Félicie Poterin du Motel, the daughter of the physician at the hostel where Galois stayed during the last months of his life. The reason for the duel is not clear, and Galois strongly believed he would not survive the duel. As a result, he spent the entire night before the duel writing and creating mathematical manuscripts, which have been considered one of the most substantial works of mathematics ever produced. The next morning on May 30, 1832, Galois was shot in the abdomen and tragically left to suffer. This brilliant mathematician died the next morning in the hospital at the age of 20, possibly of peritonitis. Galois' manuscript was published in 1846 and contained a unique proof that no general radical formula exists for a polynomial equation of degree 5 or higher.

The Norwegian mathematician Niels Henrik Abel (1802–1829) had already proved in 1824, at the young age of 22, that no general radical formula exists for a polynomial equation of degree 5. However, Galois' work went further and deeper, and could be applied to prove for any polynomial equation whether it can be solved with radicals. His research spawned an entire new field, today referred to as *Galois Theory*, which has led to simpler solutions of more advanced mathematics and has been used to show that two classic problems of geometry, trisecting an angle and doubling the volume of the cube, cannot be solved under certain given conditions. This was truly the remarkable work of a teenage genius!

UNIT 99

A Look at Einstein's Special Relativity

The theoretical physicist, Albert Einstein (1879–1955), one of the most brilliant men that ever lived, changed the world forever in 1905 when he published his Special Theory of Relativity which contradicted classical Newtonian ideas about physics. In a Newtonian world, where objects do not travel very fast, the concept of relative speed says the following. Suppose you are rowing a boat in a river at 4 mph *with* a current of 2 mph. Your speed relative to the river is 4 mph but to an observer on land your speed relative to the observer is $4 + 2 = 6$ mph. In an Einsteinian world, where objects travel very fast, close to the speed of light, which is approximately $c = 186{,}000$ miles/sec, this simple addition of relative speeds does not apply. The underlying principle of Special Relativity is that the speed of light is the same for all observers. This means that if a person is traveling extremely fast in a spaceship, say, equal to 0.70c (70% of the speed of light), and flashes a beam of light ahead, the speed of the light beam to somebody standing still will *not* be increased by the speed of the spaceship, that is, it is not c + 0.7c, but will be the same speed, c, experienced by the person in the spaceship. As strange as this seems, it has been shown by experiments to be true, and for it to be true certain extraordinary phenomena must take place.

The relationship between the speed of light c, the time t, and the distance d, is $c = \frac{d}{t}$. That is, speed equals distance divided by time. If the speed of light c is to remain the same for the person standing still and the person in the spaceship, either the distance the light travels must change, or the time elapsed must change for the person in the spaceship. Actually, according to Einstein, both the time *and* the distance experienced by the person in the spaceship change. First, *the time experienced by the person in the spaceship decreases,* so the ratio $\frac{d}{t}$, which is the speed of light, increases somewhat. This is called *time dilation.* However, the other phenomenon of Special Relativity is that the *distance experienced by the person in the spaceship decreases.* This is called *length contraction,* so the ratio $\frac{d}{t}$ remains constant. Einstein called this the *Spacetime continuum* where time is the fourth dimension. Now this can be hard to accept, and when Einstein proposed it, many experiments were subsequently performed by scientists to test it.

Here is the Hafele-Keating experiment done in October 1971, which produced conclusive results of time dilation. The American physicist Joseph C. Hafele (1933–2014) and the American astronomer Richard Keating (1941–2006) took four cesium-beam atomic clocks, accurate to a billionth of a second, on commercial airliners and traveled around the world with the clocks in opposite directions. Their time difference was compared to a clock at the United States Naval Observatory which moved with the rotation of the earth. The eastward flight traveled in the direction of the earth's rotation and therefore moved faster than the clock on earth because the rotation of the earth was added to the speed of the plane. The westward flight moved against the rotation of the earth, and therefore moved slower than the clock on earth. The eastward flight took 41.2 hours while the westward flight took 48.0 hours. The time on the eastward clock ran slower than the clock on earth because of its added speed, while the time on the westward clock ran faster than the clock on earth because of its slower speed. The time differences were very small, on the order of nanoseconds (1 ns = 10^{-9} sec), but within the predicted range of the theory of Special Relativity. The results were published in 1972 and are shown in Figure 6.35.

A noticeable effect of time dilation on how one ages may only be experienced in some future space exploration world, however astronauts on the international space station get to age just a tiny bit slower than people on Earth.

	Nanosecond difference	Nanoseconds predicted
Eastward	−59 ± 10	−40 ± 23
Westward	273 ± 7	275 ± 21

Figure 6.35

UNIT 100

A Clear Look at Calculus

The study of calculus may sound advanced to those who have never taken such a mathematics course, however it is based on a few easily understood elementary ideas. In Latin 'calculus' means pebble and because the Romans used pebbles to do addition and subtraction, one of the discoverers of Calculus, the German mathematician Gottfried Leibniz (1646–1714), coined the word for the new discipline. The origin of Calculus is a fascinating story because it was independently discovered by Leibniz and the English mathematician Isaac Newton (1643–727). Though Newton developed his critical insights during the plague years of 1665–1666, when he was sequestered in the country, it was Leibniz who published his concepts first in 1675. Consequently, there ensued a rift between England and Germany as to which native son should be credited with discovering the new mathematics and who, perhaps, stole the ideas. Today it is clear that each produced the ideas independently, however, England and Germany for over a century had little mathematical collaboration and some still do not accept the fact that calculus was independently arrived at. It was Leibniz whose notation we still use today rather than Newton's, which was more cumbersome.

Calculus deals with the infinite and the infinitesimal, and can be thought of as the mathematics of motion and constant change. It was motion, that is, the movement of objects in the earth's gravitational field, that inspired Newton to unveil the ideas of calculus. Leibniz was motivated by the change in the curve of a mathematical function and the area that its curved border encloses. We will now look more closely at these concepts to better understand what calculus is actually all about. There are two types of calculus: *differential calculus* and *integral calculus* which are essentially the mathematical inverses of each other.

Differential Calculus

Differential calculus deals with infinitesimal changes in the value of a function that changes continuously, such as distance, time, velocity, temperature, force, electrical current, etc. Consider the example of the fabricated image of the apple falling on Newton's head which supposedly motivated him to produce his laws of gravity. The velocity of the object released from some height above the ground and falling to the earth, continually increases because of the gravitational force constantly acting upon it. Over any finite period of time, you can calculate the average velocity of the object by dividing the distance traveled by the time elapsed. For example, during the first *second* the distance traveled by a free-falling object is 16 ft. The average velocity is then $\frac{16}{1} = 16$ ft/s. After two seconds, the total distance traveled is 64 ft. The average velocity during the second *second* is then calculated as follows. The distance traveled during the second *second* is $64 - 16 =$ 48 ft and the time elapsed is 1 second so the average velocity is $\frac{48}{1} = 48$ ft/s. However, what if you want to know the velocity of the object the moment it strikes the ground? This is not the average velocity but the *instantaneous velocity*, and it cannot be determined by simply dividing the distance traveled by the time elapsed because in an instant the distance traveled and the time elapsed are both zero and $\frac{0}{0}$ has no defined value. What differential calculus does to determine the instantaneous velocity, is to calculate the average velocity over smaller and smaller time intervals, closer and closer to the moment the object strikes the ground. This produces a series of values for the average velocity that can be shown to approach closer and closer to a certain limiting value, and will not exceed that limiting value. It is, therefore, deduced that this limiting value is the instantaneous velocity. If we assume the object strikes the ground at exactly 2 seconds, the table in Figure 6.36 illustrates this process, and shows the limiting value to approach 64 ft/s. The speedometer in your car is an example of instantaneous velocity and is what calculus allows us to determine for any function.

This process of finding the instantaneous change in the value of a function, in this case the distance function, is called *differentiation*. In differential calculus, differentiation leads to a formula, called the *derivative*, which can calculate the limiting value of a function directly without making a series of calculations. This example reveals that the velocity function is the derivative of the distance function because it

Time Interval (sec.)	Time Difference (sec.)	Distance Traveled (ft)	Average Velocity (ft/s)
1.5–2.00	0.5	28	56
1.9–2.00	0.1	6.24	62.4
1.95–2.00	0.05	3.16	63.2
1.99–2.00	0.01	0.6384	63.84
1.999–2.00	0.001	0.063984	63.984

Figure 6.36　Limiting value of average velocity $= 64$ ft/s

represents the instantaneous change in the distance at any moment. Similarly, an acceleration function is the derivative of the velocity function because it represents the instantaneous change in the velocity at any moment. Differential calculus studies the derivatives of functions and applies these concepts to solve many practical problems involving the rate of change of functions.

Integral Calculus

Integral calculus also concerns itself with instantaneous change, but not with a curve, but with the area bordered by a curve, which is dependent on the changes in the curve. Mathematicians struggled for many years to find formulas for the areas of curved figures by approximating them with straight lines. One of the earliest brilliant calculations was that done by Archimedes (212 BCE–287 BCE) who first approximated the area of a circle by averaging the areas of an inscribed hexagon and a circumscribed hexagon. He then increased the accuracy of the result by doubling the number of sides of the hexagons four times until they became inscribed and circumscribed polygons of 96 sides and computing their areas. He found the value of π to be between $3\frac{1}{7}$ and $3\frac{10}{71}$ or between 3.1429 and 3.1408 which was a remarkable result at the time. His calculation was based on the *Method of Exhaustion*, developed in the 5th century BCE by the Greek

Eudoxus of Cnidus, which is essentially the method of integral calculus as we will now demonstrate. Consider the basic curve of a parabola given by $y = x^2$ and the area under the curve bordered by the vertical lines $x = 0$, $x = 1$, and the x axis as shown in Figure 6.37.

The area under the parabola (or for any function) can be determined by first dividing the curve into a large number of small equal segments. On each segment, two rectangles are drawn, one below and one above the curve as shown in Figure 6.37. By averaging the total area of the rectangles below the curve with the total area of the rectangles above the curve, the approximate area under the curve can be determined. The *exact* area is calculated by continually increasing the number of rectangles on smaller and smaller segments and determining the limiting value of the result, similar to what is shown above for the velocity of the falling object. This process is called *Integration* and leads to a formula for the area under the curve of a function called the *Integral*. The limiting value for the area under the parabola in Figure 6.37 turns out to be $\frac{1}{3}$ of a square unit which is the exact value for the area.

A most remarkable result called the *Fundamental Theorem of Calculus* shows that the process of integration is the inverse of the process of differentiation, similar to division being the inverse of multiplication. Whereas the derivative of almost any function can be

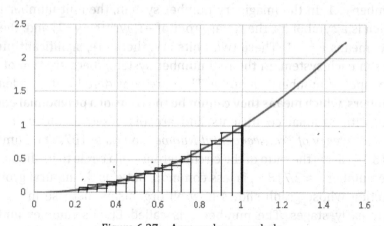

Figure 6.37   Area under a parabola

determined, it is not always possible to reverse the process and find the integral for any function. Integration is a more difficult process than differentiation and many mathematical techniques have been devised to integrate functions. The number of integral formulas that have been developed number in the hundreds. More advanced integrals enable us to also calculate volumes of figures formed from curves. Two common examples are the sphere, formed by rotating a circle about its diameter, and a paraboloid, formed by rotating a parabola about its axis. A paraboloid is a very useful figure and is the shape used for a satellite dish and a radio antenna (see the Unit on the Parabola, page 350).

Unit 101

Euler's Remarkable Relationship

In the unit *How We Categorize Numbers* on page 38, we discuss the different kinds of numbers in our system: natural numbers, integers, rational numbers, irrational numbers and imaginary numbers. The first four sets of numbers belong to the real numbers and are distinct from the imaginary numbers. The real and the imaginary number systems each contain a unit number, which is the building block for all the other numbers in the system. In the real number system, the unit number is 1. In the imaginary number system, the unit number is i, which is a symbol for the square root of -1, written $\sqrt{-1}$, and, therefore, means $i^2 = -1$. These two units are, therefore, significant numbers in each system. In the real number system, there are two other very special numbers, π and e. They are considered transcendental numbers, which means they cannot be the roots of a polynomial equation with rational coefficients and are discussed in the unit *The Transcendency of Transcendental Numbers* on page 107. The number $\pi \approx 3.1415...$ is the ratio of the circumference of a circle to its diameter. The number $e \approx 2.718...$, is less common, but e^x is the natural growth function whose graph shows how all organisms increase in size in their early stages. The number e is called Euler's number and is named after the great Swiss mathematician Leonhard Euler

(1807–1883). Euler was also an accomplished physicist, astronomer, geographer and engineer and probably wrote more about mathematics than any other mathematician in history. Among his many discoveries in mathematics is the extraordinary equation, which links the five significant numbers mentioned above: 1, i, π, e and 0:

$$e^{\pi i} + 1 = 0$$

The American physicist Richard Feynman (1918–1988) called this formula "our jewel" and "the most remarkable formula in mathematics." For one not familiar with the mathematics of imaginary numbers, this relationship may appear quite strange. How can the imaginary unit i be an exponent? Hopefully, the following will shed some light on this magical formula. In calculus, there are infinite series formulas for different functions. The one for e^x is:

$$e^x = 1 + x + \frac{x^2}{2} + \frac{x^3}{(2)(3)} + \frac{x^4}{(2)(3)(4)} + \frac{x^5}{(2)(3)(4)(5)} + \ldots$$

This formula can be used to calculate the value of e^x for any value of x, to any degree of accuracy, by adding more and more terms. For example, when $x = 1$ in this formula,

$$e^1 = e \approx 1 + 1 + \frac{1}{2} + \frac{1}{(2)(3)} + \frac{1}{(2)(3)(4)} + \frac{1}{(2)(3)(4)(5)} = 2.717$$

and this series will approach the value of e closer and closer as one calculates more and more terms.

Similarly, there are infinite series formulas for sin x and cos x, which can be used to calculate the values of sine and cosine for any angle x, to any degree of accuracy:

$$\sin x = x - \frac{x^3}{3} + \frac{x^5}{(2)(3)(4)(5)} - \ldots$$

$$\cos x = 1 - \frac{x^2}{2} + \frac{x^4}{(2)(3)(4)} - \cdots$$

The plus and minus signs alternate in the above formulas. Notice that the terms in the sin x and cos x series also appear in the series for e^x. Now what Euler did is introduce the imaginary unit into the series for e and he obtained the following series:

$$e^{ix} = 1 + ix + \frac{(ix)^2}{2} + \frac{(ix)^3}{(2)(3)} + \frac{(ix)^4}{(2)(3)(4)} + \frac{(ix)^5}{(2)(3)(4)(5)} + \cdots$$

Because $i^2 = -1$, it follows that $i^3 = -i$, $i^4 = 1$, $i^5 = i$ and so on, with the pattern repeating those four values in that order as the exponent increases. The above series then simplifies to:

$$e^{ix} = 1 + ix - \frac{x^2}{2} - \frac{ix^3}{(2)(3)} + \frac{x^4}{(2)(3)(4)} + \frac{ix^5}{(2)(3)(4)(5)} - \cdots$$

Two minuses and two pluses alternate in the above formula.

If you separate this series into two series, one containing all the terms with i and the other all the terms without i, you can express e^{ix} as follows:

$$e^{ix} = \left[1 - \frac{x^2}{2} + \frac{x^4}{(2)(3)(4)} - \cdots \right] + \left[ix - \frac{ix^3}{(2)(3)} + \frac{ix^5}{(2)(3)(4)(5)} - \cdots \right]$$

Observe that *the series in the left bracket is the same as the series above for cos x, and the series in the right bracket is the same as the series above for sin x multiplied by i.* You can therefore write a formula for e^{ix} in terms of cos x and sin x:

$$e^{ix} = \cos x + i(\sin x)$$

This is the famous formula that Euler came up with and it is called *Euler's Formula*.

And now for the final result. If $x = \pi$ in Euler's formula then:

$$e^{\pi i} = \cos \pi + i(\sin \pi)$$

Note that π radians $= 180°$ and $\cos \pi = -1$ while $\sin \pi = 0$. Substituting these values in the above result gives:

$$e^{\pi i} = -1, \text{ or } e^{\pi i} + 1 = 0$$

This is the way Euler's remarkable relationship evolves. Even if you did not grasp all of the above, you can still marvel at this incredible result relating these five universally-significant numbers: 1, i, 0, π and e.

Index

375

Printed in the United States
by Baker & Taylor Publisher Services